To Vic.
December 93.

Time in Gold
Wristwatches

Seiko Kinetic – harness the power

Seiko Kinetic. Converting your every movement into never-ending energy. The perpetual accuracy of quartz. Extreme reliability. Yesterday was battery. Today it's Kinetic – exclusively by Seiko. 15 bar water resistant. One way rotating bezel. Built to last. Challenge the elements. **The time is now.**

Ah, the thrum of the road. The wind in your hair. The sheer, glorious, hang-in-there intoxication of open-top motoring.

Unless, that is, you happen to be driving our new Golf Cabriolet.

A soft, gentle breeze is the most you can hope for.

And scarcely that with the optional anti-draught screen clipped onto the roll-bar behind you.

As for ragged surfaces and six-inch pot-holes, the Golf, we confess, fairly soaks them up.

Killjoys that we are, we've reinforced the floorpan. Strengthened the front and rear bulkheads. Stiffened the door sills.

The Spirit of Time

CYMA
Since 1862

Featured, mens solid steel bi-colour bracelet model 02.001.003 £149.00.
Also in gold plated 02.0001. 001 £170.00 Both Models available in ladies sizes.

Further information contact Stelux Watch UK Limited, Stelux House, Lichfield, Staffordshire, WS13 6PW.
Telephone: 0543 414211

13mg TAR (
SMOK
Health Department

DESIGNED to PERFORM...

...anytime...anywhere

'Liberté 420' series, water resistant to 100 metres. Stunning good looks, whether dining, dancing, swimming or sailing; work or play, 'Liberté' is a perfect balance of strength and active style. The watch features a Swiss precision movement, solid bracelet with concealed clasp, and date.

Available from Harrods Fine Watch Department, Selfridges, selected branches of Goldsmiths, Walker & Hall, Watches of Switzerland, Beaverbrooks, John Lewis Partnership, F. Hinds, Emson Haig at Lakeside and other leading Jewellers throughout the United Kingdom and Ireland.
For further information Telephone: 0543 414211.
Models featured Gents £225.00, Ladies £245.00.
Also in Stainless Steel £185.00.

ellesse watch

ANTAEUS
POUR HOMME

DE TOILETTE

CHANEL
PARIS

SECTOR SPORT WATCHES. UNLIMITED TECHNOLOGY FOR UNLIMITED FEATS.

Sector Sports Watches and the No Limits team. Technology and incredible sporting feats linked by a single end: to go further, to go beyond all limits. To believe in something, then to do it. In fact our watches are constantly being taken to the limit by athletes such as Patrick de Gayardon in extreme parachuting and base jumping, by the extreme canoeist Shaun Baker as he takes on the most awesome of waterfalls, by the free-diver Umberto Pelizzari and many others. Now you too can get a piece of the No Limits action by buying a Sector Sports Watch. And what's more, with every purchase, Sector and the Official Sector Stockists are proud to offer you the extraordinary "No Limits Team" book of action photography. So now you know, why limit yourself to a normal watch.

ADV 4500 CHRONO

Sector Sport Watches from £125,00 to £850,00. Available from Harrods Fine Watch Department, The International Watch Room at Selfridges, Selected Branches of Watches of Switzerland, Zeus, Mappin & Webb, Ernest Jones, Leslie Davis, John Lewis Partnership and other Leader Jewellers throughout the United Kingdom and Ireland. For further information please call telephone 0703 271233.

THE NEW DYNAX 500si. AN SLR SO ADVANCED IT'S EVEN GOT YOUR FAVOURITE PROGRAM BUILT IN.

 The Clothes Show.

With the Dynax 500si, you can pick exactly the right program, or mode, to suit the shot.

Say you're taking a picture of your other half in their latest creation.

The only subject you'll want to see is him or her, not next door's washing or any other distracting background detail.

Choose the Portrait Mode and the 500si really makes your subject stand out by making the background blurred.

 Wish you were here.

There are occasions when the background can be as important as the people you're taking.

Maybe to prove you really did visit some exotic faraway spot.

The Landscape Mode enables you to make sure that everyone and everything is clearly in the picture. Clearly being the appropriate word.

This is because the Landscape Mode actually adjusts the exposure to ensure that both subject and background are as sharp as possible.

 Gardeners World.

Some of the most beautiful sights on earth may be no further than your garden.

And they're yours to capture by switching to the Close-Up Mode on the 500si.

Flowers and small insects take on a three-dimensional quality when you take pictures using this program - because the subject is shown in sharp relief against a background that's automatically blurred. Meaning gardeners' question time now becomes a thing of the past.

Match Of The Day.

How often has the action in front of you gone too fast for your camera to keep up with?

Well not any more. With the 500si you can be certain of keeping up with the pace of play.

Simply switch to the Sports Action Mode and you're all set to create impressive fast-action pictures every time.

It works like a dream because the camera keeps focusing continuously whilst selecting the fastest possible shutter speed.

News At Ten.

One of the best programs is on at night. It's called the Night Portrait Mode and once you're switched on to it, you're all set to shoot the night away.

It allows you, for example, to take amazing portrait shots in the evening - with both subject and background receiving equally dramatic emphasis. Or, if you prefer, stunning night scenes.

Either way, it's one program you won't want to miss.

THE CLOTHES SHOW.

WISH YOU WERE HERE.

GARDENERS WORLD.

MATCH OF THE DAY.

NEWS AT TEN.

The Dynax 500si allows you to be as creative as you want to be.

You can for example, simply switch to Program Mode then point and shoot.

The pin-sharp automatic focus closely monitors your subject, determining whether or not it's moving or stationary, right up to the moment you press the button.

And the unique exposure metering system actually measures exposure levels in not one, but eight separate areas of the picture to give perfectly balanced results.

The Subject Program Selector (described left) is a step up from the 'P' Mode. Because you can choose the mode to fit the subject.

Finally, by selecting Aperture Control, Shutter Priority or Manual, you can experience full creative control through the viewfinder.

With its ability to combat 'Red Eye', through its dedicated wireless off-camera flash guns, the 500si is a very special SLR indeed. And it's available with an equally **MINOLTA** impressive range of top quality lenses and accessories.

Ring the Minolta information line on Freephone 0800 622 283 for further details delivered to your door and you'll soon be tuning into your favourite program.

Minolta Dynax 500si kits containing camera, zoom lens, gadget bag, wide strap, extra battery, photographic book and films: with 35-70 lens £400.00, with 28-80 lens £470.00.

culture

YEAH, BUT HOW DO YOU REALLY FEEL?
According to a Gallup Poll, March, 1993:
42% of Americans own a gun.
57% of gun owners favour stricter American gun laws.
36% of gun owners want no change in gun laws.
82% of non-gun-owning Americans favour stricter American gun laws.
13% of non-gun-owning Americans want no change in gun laws.
60% favour a total ban on the possession of assault weapons.
60% favour limiting individuals to one gun purchase per month.

KILLERS IN THE HOME
27% of gun owners have one gun in the house.
22% have two guns.
14% have three guns.
31% have four guns or more.
6% are not sure how many guns are in the house.
25% of American households have a handgun

within their walls.
45% of American households have a gun of some type within their walls.
90% say they know how to use their gun.
78% have fired their gun.
43% say their gun is loaded right now.
90% know where their gun is right now.
39% carry a gun in their car or truck.
24% carry a gun on their person.

8% claim to have used or threatened to use their gun to protect themselves or their property.
16% say they have been threatened by someone with a handgun.
77% believe their right to own a gun is guaranteed by the constitution.
43 The number of times more likely it is that a gun kept in the home will kill a family member or acquaintance than to kill an intruder.

SUBURBAN ASSAULT
Kids and Guns
14 The number of children aged 19 and under killed in America per day by guns.
4% of high school children carried a gun during a one-month period in 1990.
100% Increase of gunshot wounds to children under sixteen between 1987 and 1990.
77% Increase in firearm death rate among teens from 1985 to 1990.
105.3 Number of firearm homicide deaths per 100,000 black teens aged 15-19.
9.7 Same number for white teens aged 15-19.
300% Increase of black male teenage homicide rate from 1985-1990.
100% Increase in juveniles charged with homicide from 1982-1990 (CDC).

MORE DEATHS
43,536 Americans killed in motor vechicle accidents in 1991. This is the most likely way you will die in America.
38,317 Americans killed from gunshot wounds in 1991. Second most likely way you will die in America.
1995 or 96 The year in which firearm deaths should surpass motor vechicle deaths in America. (Center for Disease Control Estimate)
28.4 per 100,000 white Americans die from motor vehicle injuries.
15.2 per 100,000 from firearms.
28.7 Rate for Hispanics – motor vehicles.
29.6 Hispanics – firearms.
23.0 Rate for blacks – motor vehicles.
70.7 Blacks – firearms.
13.7 Rate for Asian/Pacific Islanders – motor vehicles.
10.7 Asian/Pacific Islanders – firearms.
60% of deaths among black teenage males (15-19) resulted from a firearm injury.
23% of deaths among white teenage males resulted from a firearm injury.

HOMICIDES
12.9 out of 100,000 white males aged 20-24 were killed by firearms in 1990.
140.7 out of 100,000 black males of the same age were killed by firearms in 1990.
1 in 28 black males born in the US is likely to be murdered.
93% of black murder victims were slain by black offenders. ⓢ **CHRIS DIXON**

▶ JUST SIGN ON THE DOTTED LINE…TO PURCHASE A HANDGUN IN AMERICA (LEGALLY)
21 Age Requirement.
No Answer to the question: "Are you a convicted felon?"
No Answer to the question: "Are you a drug/alcohol abuser?"
No Answer to the question: "Ever been adjudicated mentally ill?"
No Does the Federal Government check the accuracy of the answers?
16% Criminals who obtain their handguns legally through gun dealers.
90% Handguns used in crime in New York City which are bought out of state. (New York has tough gun purchasing laws)
87% Handguns used in crime in Texas which are bought in Texas. (Texas has very weak gun purchasing laws)
3 The number of states which require a safety training course for gun buyers.
3 The number of states which require guns or gun owners to be registered.
34 Number of convicted murderers in California who were stopped from purchasing guns in 1991 because of the state's 15-day waiting period. (Independent State Law)
$10 The amount it costs to purchase a Federal License to sell guns.
1,509 Average number of gun vendor license requests received by the Bureau of Alcohol, Tobacco & Firearms every month.

REMEMBER THE ALAMO
This is the arsenal of the Branch Davidians – the Texas cult led by David Koresh whose compound was burned to the ground in late 1993 with the loss of 85 lives. The following were all purchased legally.

- 123 AR-15/M-16 type assault rifles.
- 100 fully automatic AR-15/M-15 machine-guns.
- 44 AK-47 type assault rifles.
- 2 "Street Sweeper" assault shotguns (the gun preferred by Arnold Schwarzenegger in *The Terminator*).
- 2 Barrett .50 calibre anti-tank rifles.
- 26 M-1 military rifles.
- 11 other long guns.
- 60 handguns, among them MAC 10 and MAC 11 semi-automatic pistols.
- 20 100-round capacity drum magazines for AK-47 rifles.
- 260 AR-15/M-16 magazines (which hold 20-30 rounds apiece).
- 500,000 rounds of ammunition.
- Numerous: silencers, machine-gun conversation parts, homemade grenades.

With 800 matches in just two weeks we can't afford to lose a second.

The greatest of the world's players build their game on accuracy, consistency and strength.

As timekeepers to the Wimbledon fortnight, we build our success on the very same principle.

We have to, for the vagaries of the English summer are such that it is by no means unusual to see a tarpaulin appear on the Centre Court instead of the world's premier players.

Perhaps it is not surprising, then, that the organisers of the Wimbledon fortnight have come to rely on Rolex of Geneva; with so many matches to be played in so very few days, it is crucial that the event runs according to schedule.

Rolex clocks placed throughout the grounds keep the public aware of the exact time.

Nor indeed has the Rolex reputation for accuracy been lost on the players, for a Rolex is perfectly suited to meet the demands of a life on the professional tennis circuit.

Its case, for example, is formed from a single block of metal. This gives it a strength which will preserve the accuracy of a movement that has taken up to a year to create.

Such strength is more than a match for the high-speed services, lobs and smashes of powerful players such as Jim Courier. For as they and Rolex both know, in tennis, perfect timing is an incalculable advantage.

ROLEX
of Geneva

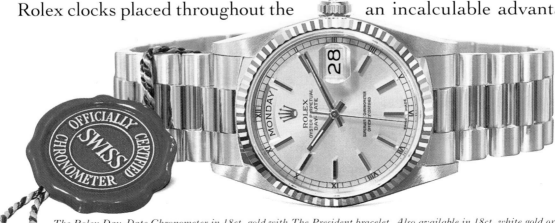

The Rolex Day-Date Chronometer in 18ct. gold with The President bracelet. Also available in 18ct. white gold or in platinum.

Only a select group of jewellers sell Rolex watches. For the address of your nearest Rolex jeweller, and for further information on the complete range of Rolex watches, write to: The Rolex Watch Company Limited, 3 Stratford Place, London W1N 0ER or telephone 071-629 5071.

**APPEARANCES CAN BE DECEPTIVE.
SOMETIMES.**

nd the R1100GS is just the motorcycle to get you there.

With its adjustable windshield and seat height, and carrying capacity that can take the full complement of BMW luggage, two people can travel in comfort.

So, in the case of the R1100GS, the camera doesn't lie. It really is a motorcycle to overshadow everything else on the road.

THE NEW BMW R1100GS

Gerald Viola * Gisbert L. Brunner

TIME IN GOLD

WRISTWATCHES

1469 Morstein Road, West Chester, Pennsylvania 19380

English edition copyright © 1988 by Schiffer Publishing Ltd.
Translated from German by Dr. Edward Force
Originally published in German as *Zeit in Gold, Armbanduhren*, by
Signum Medien Verlag. München
Library of Congress Catalog Number: 88-61466.

All rights reserved. No part of this work may be reproduced or used
in any forms or by any means—graphic, electronic or mechanical,
including photocopying or information storage and retrieval
systems—without written permission from the copyright holder.

Printed in the United States of America.
ISBN: 0-88740-137-6
Published by Schiffer Publishing Ltd.
1469 Morstein Road, West Chester, Pennsylvania 19380

This book may be purchased from the publisher.
Please include $2.00 postage.
Try your bookstore first.

CONTENTS

Foreword	6
Introduction	8
Audemars Piguet	12
Baume & Mercier	34
Blancpain	40
Breguet	50
Cartier	60
Chopard	72
Corum	82
Ebel	90
Gérald Genta	98
Girard-Perregaux	104
IWC	112
Jaeger-Le Coultre	124
Patek Philippe	136
Piaget	178
Rolex	190
Ulysse Nardin	214
Vacheron Constantin	220
Technical Terms	240
Credits	254

FOREWORD

Many years ago I fulfilled a long-standing wish and bought myself a wristwatch by Patek Philippe, the "Golden Ellipse" model. That was in 1975, at a time when wristwatches had not yet been found to be collectible. The fact that it was a watch from that firm is related to my inheriting the obligation. In 1914 my grandfather had founded a journal about watches. Thus my fascination with wristwatches developed very early in life.

So I finally possessed such a costly piece, worth a small fortune, and my pride of possession made me curious to learn more about this tradition-rich firm than I could find in the flowery descriptions in their colorful pamphlet. But my efforts went for naught. I could not learn anything concrete except a few scraps of information. I would gladly have read a book about the history of Patek Philippe, my watch. I still had to be patient for many years, until in 1982 a wonderful Patek Philippe book appeared. Unfortunately, it only partially satisfied my craving for knowledge, since the book dealt mainly with pocket watches, to which I felt no strong attraction. My fascination was and is with the wristwatch in its highest state of technical refinement, as a mechanical masterpiece in our times of quartz and computers.

A small collection of various brand-name wristwatches of present-day production gave the impetus, for there was little information to be had about other distinguished Swiss brands either. The decision to write a book about Swiss luxury wristwatches was born.

Full of inspiration, I traveled to Geneva in February of 1985, to the Mecca of Swiss watch manufacturers. I visited several of them and expressed my idea of writing a book about them. The results of that trip ranged from detachment to rejection of my plan. Despite this, though, I did not let myself be turned away from my project.

It was lucky that I happened to meet several people who considered my idea worthy of encouragement. I owe it to them that, through their intervention, the doors were opened behind which I thought I would discover secrets.

I was able to persuade a good friend, a journalist on the daily press, to research and write the chapters. Many conversations took place in which I conveyed to him my idea of what I wanted the book to be. It was to be a chronological historical documentation of the sixteen leading Swiss luxury watch manufacturers, it was to have fairy-tale characteristics, and it was to share the magic that everyone senses who calls a luxury watch his own.

A second author was roped in for specialized advice, technical descriptions, comprehensible explanations of the many complexities in watch construction, and for putting watches in chronological order.

I believe the book succeeded in those areas.

In the individual chapters the histories of the most outstanding watchmakers in the world are related. The beginnings of these firms go back more than two hundred years into the past. In words and pictures, history comes alive again. Emperors, kings, princes and noblemen first made possible the development of the watch into an ornament to be worn on a golden chain, or even on the arm. Their refined taste for the beautiful, the precious, and their admiration for great handworking skill encouraged the watchmakers to build the most complex watches in their small workshops, and to seek out ever-greater complexities for watches. Recognition for this was not denied them, assuring the continuity of their production to our own times.

Pictures of watches from their founders' times were available from very few firms. But the illustrations of wristwatches made at the present time show in a very impressive way what arts are still being promoted today, almost unchanged, and what brilliant watchmakers are still active today, for many of the watches shown here will one day be desirable rarities, just as wristwatches from the times before 1960 are today.

The artisans' guild of Swiss watchmakers has been able to maintain its traditions over the centuries, and to expand its significance into one of the most important economic factors of Switzerland. Despite changing fashions and quartz watches, the real masters of their trade will never die out, as this book will show.

The criteria for choosing what to include in this book have been subjective, just as the definition of "luxury" is always derived from the viewpoint of the observer. But one thing all of the manufacturers treated here have in common is that they regard themselves as the builders of watches.

During the three years of work on this book, I found many new friends, all of whom are linked in some special way with the subject of wristwatches. They are collectors of these timeless works of art, authors, dealers, auctioneers or people who simply like valuable watches. All of them continually encouraged me to finish this book when at times my courage had forsaken me because the problems never seemed to end.

I would like in particular to express my hearty thanks to the following people: Jean-Claude Biver, René Bannwart, Gérald Genta, André Goy, Martin Huber, Christian Pfeiffer-Belli, Osvaldo Patrizzi, Jacques Piguet, Christian Piaget, Karl-Friedrich Scheufele, Rolf W. Schnyder, Paul Stuber and Dr. Walter G. Tobler.

The photographic material showing present-day products was kindly provided for publication by the manufacturers. Collectors and auctioneers made available the pictures of antique watches. They too deserve my special thanks at this point.

I wish you much pleasure in reading and looking.

Munich, May 1988.
Wolfgang K. Fulde
Editor

INTRODUCTION

The phenomenon of time has accompanied and occupied the human race since its beginnings. Only the world of the small child, who begins by exploring and experiencing the three dimensions of space, is timeless. Only in one's third year of life does time push into a human being's consciousness.

What, then, is time?

Time is science.

Without the factor of time, modern natural science would be impossible. Just as one cannot describe the speed of his car without applying the dimension of time as a comparison factor, so too would planning a trip or a flight to the moon be impossible without exact timing.

The Italian mathematician and natural scientist Galileo Galilei was the first to bring the factor of time into physical formulas as an independent variable. This revolutionized the manner of scientific thinking in the Seventeenth Century.

Isaac Newton continued this realization: "Absolute time flows uniformly in and of itself, without reference to any external object."

The genius Albert Einstein, only in our century, unmasked the concept of absolute time as a contradiction in cosmic events, and anchored time in the theory of relativity as a concept of a four-dimensional, bent world.

Time is philosophy.

"The number of movements in regard to the earlier and later," is how Aristotle defined time—similarly to Plato—whereupon the thinkers thought first of all of the movements of the heavenly bodies.

In the fifth century, the church father Augustine described the past, present and future as "a kind of trinity of the soul".

And in 1781 Immanuel Kant united philosophy and physics when he recognized that "time is not a general concept, but rather a pure form of sensory perception. For rational reasons, it is the formal requirement of absolutely all phenomena.

Time is feeling.

It is only indirectly experienced by man through the events that are reflected in it. Man can see, hear, speak. But he cannot perceive time in and of itself—only feel it in relation to the events of his life space.

Time is art.

Man's ability to comprehend time, independent of external conditions (such as the position of the sun), brought the fourth dimension of time into art by means of the clock.

Man did not limit himself to the simple, technical portrayal and measurement of the time factor. In his world of personal experience he sensed the span between birth and death as something very valuable. At first he decorated the dials and hands, later even the essentially unseen technical part of the work of art, the clock.

Thus the timepiece did not remain a merely utilitarian object for telling what hour had struck. It was developed further by artisan and artist alike into a new dimension of art.

Time in gold.

How valuable time is to man is demonstrated by the integration of that noblest of all metals into watchmaking: gold. Mystic legends have always gathered around this metal, which has been the measure of economic wealth since the beginnings of civilization. In the past, goods were bought with pure gold, and over the course of history, despite all fluctuations, this metal has always maintained a high value. One could buy anything with gold—even when normal means of payment had lost their value. Until just a few years ago, the dollar, currency of the world, oriented itself to the price of gold. Metal bars by the ton were and are stored in the legendary Fort Knox to maintain the illusion that the American dollar was worth its gold.

The mystical significance of gold can be seen in the fact that nearly all religions and cults have used golden vessels or display objects.

And in all its appearances in the world of fashion, one factor has remained constant: the wristwatch that is regarded as classic is made of gold.

The watch is seen in this unique light in human life because it unites science, philosophy, emotion and art.

And it has one more quality that is unique: in the development of time measurement one can see the development of humanity, as if on an oversize dial.

The peoples of antiquity could really know nothing of this development when a clever fellow "stumbled" over his shadow one day. He noticed that his own body's shadow was longer in the morning than at midday and then grew steadily until sunset.

The correct understanding of this at first inexplicable phenomenon was gained by this unknown thinker when he stuck a stick in the ground and realized that the not particularly well-defined shadow changed not only its length but also its direction.

Until this essential fact was discovered, the fourth dimension's smallest terms consisted for man merely of the alternation of day and night. The discovery was not yet needed in its cosmic relation to the heavenly bodies.

And so the first sundial was "only" a refinement of time measurement, which until then could be divided into years by the position of the sun and into months by the appearance of the moon.

It was probably the Babylonians who—as Herodotus reports—first developed time measurement on the basis of this knowledge. There are no earlier drawings of sundials, and the only limiting factor of this kind of clock is that the life-giving rays of the earth's nearest star are prevented from casting shadows only now and then by dark rain clouds.

The actual development of modern clock technology has taken place in the last seven centuries. And when one looks at the crude old tower clocks of bygone days in the clock museum of La Chaux-de-Fonds, in Switzerland, today, one must realize that clocks were always the most complex machines of their epoch, documents of the technical high points of any era—to the atomic clocks of the present.

It was a development that ran parallel, for in the last centuries progress always demanded more and more precise timepieces.

For example, the English Parliament influenced and guided clock technology—at a time when the British were the dominant seafaring nation of the world. But—their captains had difficulties with navigation, which resulted in considerable financial losses because of shipwrecks. Seafarers could determine their position exactly by the position of the sun only when they knew the exact time. But the clocks of those days could not withstand either the ships' movements or the changes of climate. The inventors had to take a hand. And the English Parliament offered a prize for the development of the first reliable marine chronometer.

Another example: the Industrial Revolution changed the living and working habits of man. While the farmer could measure the worth of his work by the selling price of his harvest, only time could serve as the essential standard of office or factory work's achievement and its reward. As the standard of time measurement, the clock became the determining factor of modern life.

And a third example: to make space exploration possible at all, time measurement had to be made independent of the movement of the stars, which are simply too inexact for ultra-precise measurement of time. That meant that traditional time measurement on the basis of the earth's rotation and movement had to be corrected more and more often by adjusting the passage of time.

The time measurement that developed out of the phenomena of Nature had overtaken nature.

Time is precious.

Our first reference should not be to the expression "time is money", but to the fact that, when the watchmaker's art began, the possession of a timepiece was limited to the rich and privileged. Only they could afford these works of art. The tower clocks of the Middle Ages can be regarded as a social achievement.

And today?

Today time rules our lives. Our work, and thereby our income, are made measurable by time. Whoever can sell his working time dearly achieves prosperity. Time has indeed become money.

Only one thing has not changed: time is still a privilege of the successful and affluent—when it is a matter of luxury wristwatches.

These artistic treasures of the watchmaker's venerable art have their price today too. For—just as only a few such watches formerly existed—such exclusive pieces are still made by the traditional watchmaking art by only a few manufacturers at present.

Time is money.

The buyer of a luxury wristwatch notices this when he has to pay for the production time of a luxury wristwatch, which often takes more than a year.

Should he ever have his watch opened by a specialist, he can take a fascinating look at man's technical development. But a look into the heart of a timepiece cannot tell him the many stories that lie hidden behind this development. The secrets of luxury wristwatches are preserved and protected by their few exclusive producers. They maintain an absolute silence about their production methods, a silence no less complete than that of the Catholic Church's confessional.

To preserve their firms' history they maintain museums that no mere mortal will ever see from inside—even if he heard of their existence.

These manufacturers maintain absolute discretion that has put pearls of sweat on the brow, and the color of anger on the cheeks, of many a finance official. Naturally it is forgivable, for the names of the customers become known—if at all—when their mortal remains are laid to rest in some prominent graveyard.

Patek Philippe, for example, keeps its "Golden Book" (actually bound in green leather) in Geneva. For over a century, the names of famous clients have been collected in it.

But officially, only six persons are named: Arthur Rubinstein, Ella Fitzgerald, Albert Einstein, Peter Tchaikowsky, the Duke of Windsor and Walt Disney.

And with all this dealing in secrets, it is also clear that when new technical developments are achieved, the patent applications are turned in with pride but also with a soft gritting of teeth. For once again a hitherto well-kept secret has been made available to the non-chosen and thereby "desanctified".

The many legends and numerous anecdotes that collect around the exclusive timepieces of the luxury manufacturers arose by themselves—as always happens when people don't know the exact truth; assumptions are made that suddenly become "truth" because nobody can test the real truth of them.

Only with a lot of time and trouble is it possible to separate the legends from the reality. And sometimes it is simply impossible, for many handed-down legends are passed on as truth by the present-day owners of the products.

The past history of luxury wristwatches, whose story this book tells, is fascinating, adventurous, and often downright fabulous.

AUDEMARS PIGUET

Three great English warships with the name "Royal Oak"—taken from that legendary "royal oak" that is said to have offered King Charles II a safe hiding place from his pursuers—were the godparents of a watch model name: the *Royal Oak*.

With its often-copied but never equaled bull's-eye design, Audemars Piguet left its mark on watch history. The steel watch, in the highest price range, became accepted in fine salons overnight. The manufacturers had

created the first exclusive sport watch, which was as wearable at tennis, golf or polo as at an evening reception. For almost two decades this watch has been the leading model of the firm in Le Brassus, and has inspired something unique among those who know and love watches: the establishment of a club of Royal Oak wearers. Just as the familiar oak tree was raised to honors by kingly power, Audemars Piguet helped the steel used in the production of this watch to attain respect.

But today's fairytale atmosphere found in the watches that come from Audemars Piguet was not present when the twenty-three-year-old Jules Audemars started an independent business in Gimel in 1874. Watchmaking training with various great masters of the times preceded this step. After a year and a half he balanced his books, was dissatisfied with what he had accomplished thus far, and climbed back over the nearly 1500-meter-high mountain chain to his home town of Le Brassus, barely ten kilometers away. There the highly talented watchmaker Jules Audemars quickly found a new occupation: he produced raw components for watch mechanisms.

At just the same time, Edward-August Piguet, two years younger, was looking for work. Like Jules Audemars, he had learned the watchmaker's trade after finishing public school in Le Brassus. Like his future business partner, he had followed the path of training to various masters' workshops in the Vallée de Joux before finishing his studies with a course for testers taught by the president of the Joux Valley guild, Charles Capt. Now he wanted to get a job as a "repasseur" in the valley.

The meeting of the two talented watchmakers in 1875 introduced one of the most significant chapters in the history of Swiss watches. Jules Audemars and Eduard Piguet decided to master their professional fate together. That is why 1875 is accepted as the year the firm was founded. At first they looked for orders, built watches, but also bought parts from other specialists. Even then a division of work could be seen. Jules Audemars took over the technical side, while Edward Piguet concerned himself with finances and sales. After several

Jules Audemars
Edward Piguet

Title page of a joint catalog of Audemars Piguet and the Roy dealership of Paris around the turn of the century. Shown is a gold Savonette cased watch with grande and petite sonnerie, minute repetition, perpetual calendar, moon phase and age, chronograph with sixty-minute and twelve-hour registers, winding indication as well as a bimetallic balance.

years Jules Audemars and Edward Piguet traveled to Bern to visit the "Administration Office Technique d'Edition et Publicité" on St. Nicholas' Day, 1882, and register the trade mark that was to be engraved on the watch movements and cases from their workshop: "Audemars, Piguet et Cie." At this point the name of the firm first appeared in official documents. The official founding of "Audemars, Piguet et Cie.—Manufacture d'Horlogerie compliquée de haute précision" took place in 1889. The chosen subtitle in the firm's name shows that the two watchmakers, now 38 and 36 years old, had set the goal for the further development of the undertaking: building complex, ultra-precise watches.

In the past century there were already agencies, industrial organizations and statisticians in Switzerland. And one of them reported to posterity that Audemars, Piguet et Cie., in the year of their entry into the official register as a manufacturer, employed ten workmen and—what was not at all to be taken for granted—kept them busy all year. "AP"—as the firm is known for short nowadays in Switzerland—had already grown to be the third largest employer for watch manufacturing in the Canton of Vaud.

Soon after the founding of the company, Jules Audemars and Edward Piguet opened a branch office in the watch center, Geneva. They decided not to limit themselves to producing individual components and assembling watches, but to do the complete job themselves. That was a radical break with tradition, but it led to the end result that the products they manufactured in Le Brassus were more precise and perfected than those of many competitors.

A series of watches from this era did not bear the trade mark registered in Bern, for many of the retailers wanted to buy watches and put their own signatures on them. Nevertheless, only products of high and highest quality left the workshop. Very soon the greatest, most renowned retail houses were among those ordering watches from Le Brassus. Complete watches of every kind, with or without complication, with no identifying mark of Audemars, Piguet et Cie. to be found on either the plate or the dial. Watches still exist today that are signed "Tiffany & Co." or "Van Cleef & Arpels", and that keep many a collector in the dark as to the true origin of their treasures.

Pages from an advertisement of the firm, circa 1930:

(upper left) Cushion-shaped wristwatch with chronograph and 30-minute register; start-stop-return of the chronograph are controlled via the winding crown.

(lower left) Rectangular wristwatch with complete simple calendar and moon-phase indication.

(upper right) Pocket watch with chronograph and 30-minute register.

(lower right) Pocket watch with perpetual calendar and moon-phase indication.

The United States were an important market for Audemars Piguet from the beginning, although Jules Audemars made the long sea voyage across the Atlantic only twice in his life, and high customs duties made trade difficult. At that time the Americans levied very high import duties on finished watches. The Swiss manufacturers—not just Audemars Piguet—got around such limitations by delivering only the movements. In the New World these masterful movements were built into the importers' cases, or into cases made according to the original plans supplied from Switzerland. The production numbers from this era are interesting. They were engraved consecutively in the watches, irrespective of calibers or other differences. These numbers were used since 1892. The first watch sold in

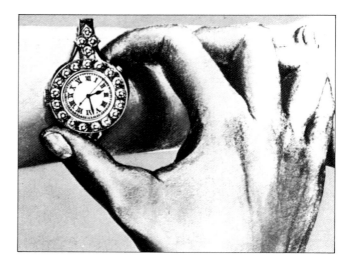

that year was number 4341, the last 4505, making exactly 165 timepieces for 1892. Further production figures to the turn of the century show, to be sure, that the increase in production was influenced strongly by favorable trends and other uncontrollable factors,

Page from a 1930 advertisement, showing men's rectangular wristwatches.

Woman's diamond-set wristwatch with minute repeat and sweep seconds; movement no. 13931, finished in 1911.

Advertisement in the Swiss "Golden book of Watchmaking", 1927, showing:
Upper left: Formal watch with digital date, day and month indication, but no perpetual calendar.

Lower left: Gold-enamel pocket watch in art-deco style with minute repeat, perpetual calendar, moon-phase indication and chronograph.
Upper right: Man's wristwatch with digital hour and minute indication.

Lower right: rectangular wristwatch with large moon-phase indication.

but the next business year, 1893, was recorded in the annals of "AP" as the best before the turn of the century: 691 watches were produced! The production figures for the following years:
— 1894: 340 watches,
— 1895: 194 watches,
— 1896: 153 watches,
— 1897: 217 watches,
— 1898: 114 watches, and
— 1899: 190 watches.

The extent of Audemars Piguet's mastery of technical development as of 1890 is shown by a gold pocket watch which features a chronograph with 60-minute register and an "independent second hand". The masterpiece is also equipped with a "seconde foudroyante". This little "lightning second hand" makes one turn per second, divided into four microsecond "jumps". The movement has two barrels, and the wheel train is made for 14,400 beats of the balance per hour. A further example of the highly developed products being manufactured is surely an open pocket watch of 1894 which, in addition to its chronograph with sweep hand and 30-minute register, also offers minute repeat. The development of the most complex models was advanced unerringly by Jules Audemars, who had become the owner of numerous patents over the years. One example might be mentioned here, since it best reflects the philosophy of the firm, and since the watch, refined over and over technically, is built today almost unchanged. Only about a hundred examples have left the house of Audemars Piguet to this day. This shows the complexity of the watch, as well as the time of almost a year required to complete it. The present price of almost 400,000 German marks seems to play only a secondary role. It cannot be denied that this masterpiece is presently the most expensive series-production watch in the world. It is the "Grande Complication", the prototype of which was introduced by Jules Audemars and Edward Piguet at the 1889 World's Fair. This pocket watch offers remarkable things in addition to normal time indication: minute repetition, perpetual calendar and chronograph.

At this point we can allow ourselves a small jump into the present. For the task of preparing the more than 400 components that

Woman's watch with minute repeat, usable as either a wristwatch or a pocket watch; striking control by pressure on the diamond located in the case rim near the 6; 1920.

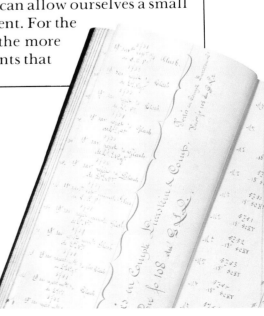

A look into the firm's archive book.

are needed and assembling them into a functioning watch at Audemars Piguet, only one man, with his workmen, has been responsible for the past fifteen years: Michel Rochat, watchmaker and artist in his specialty. Only after working thirty years as a watchmaker for this firm was he thoroughly trained in this area by his predecessor Simond, before the manufacturers turned this task, probably their most difficult, over to him. At this time the artist has another responsible task: he is training his successor.

After this side trip we return to the year 1889, when Jules Audemars and Edward Piguet laid the foundation for the heritage of their heirs with the development of this watch.

A renowned manufacturer naturally cannot neglect to offer his customers a well-conceived service plan. Audemars Piguet turned their retail houses in Geneva and London into genuine branch offices, in which watches were not only serviced but also assembled. Thus the firm now had three production facilities.

Audemars Piguet was already experimenting with wristwatches at an early date:
- 1906: a wristwatch with minute repeat, for example, signed "Gubelin" but with a movement by Audemars Piguet in a case of eighteen-karat gold (28 x 28 mm), sold in 1908.
- 1909: a wristwatch with minute repeat, with an 11.5-ligne lever movement, caliber SMV, 29 jewels, flat hairspring, cushion-shaped platinum case (30 mm diameter), sold only in 1925 to Metric Watch in New York, at that time the American importer of Audemars Piguet.
- 1911: a woman's small wristwatch with a 10-ligne movement, with minute repeat and center seconds. This watch was sold to a Berlin firm two years later.

Without a doubt, these strivings centered around the pocket watch during the founders' lifetimes. Again and again they showed that even the "farthest-out" indications could be put in a watch.
- An open pocket watch (shown at the 1915 Geneva Clock Fair) with two dials and the following functions: grande and petite sonnerie (with strike-silent control) and minute repeat striking, indication of the mean solar time on one of the two large dials, with one revolution of the large hour hand in twelve hours, indication of equation, of true and mean solar time on

Wristwatches, circa 1960.

AUDEMARS PIGUET 17

The model "Grande Complication" with a retail price of about 400,000 marks, the most expensive series-production watch in the world. Only one of these pocket watches with minute repeat, perpetual calendar, moonphase and chronograph, is made per year by Audemars Piguet.

two separate twelve-hour dials, moon age and phases, temperature indication from a bimetallic balance, seconds bit, 24-hour dial, zodiac sign, weekday, date, perpetual calendar including leap year, winding indication up to 32 hours, indication of the current year by five dials, one above the other, lasting for ten years, indication of the seasons, summer and winter solstice and day-night equation.

— A pocket watch with two dials, one on each side. The times for two different time zones can be set by the crown. The watch also has minute repeat striking, an alarm with coiled gongs, a chronograph, a barometer and altimeter to 3000 meters, and a compass. The whole mechanism consists of three different wheel trains combined in one case.

— The masterpiece of the two Le Brassus watchmakers, though, was the two-sided pocket watch with the production number 16869, which was ordered by the English importer Guignard & Golay in 1914 but only finished and delivered in 1921, after the firm founders' deaths. An English-type 26-ligne movement, one-minute tourbillon, grande sonnerie, chronograph with minute and hour counters, perpetual calendar with indications springing at midnight. Moon-phase equation, winding indication. On the back is an additional 24-hour indication with an hour hand that can be set separately (for star time) and a special apparatus that allows the position of the heavens over London at any day or night hour to be read through a window in the dial (driven by the wheel train for star-time indication). The watch's sky consists of a total of 315 stars, portraying the following constellations: Perseus, Andromeda, Cassiopeia, Cepheus, Leda, Pegasus, Draco, Lyra, Serpens, Taurus, Aries, Ursa Major and Minor, Hercules, Virgo, Hydra, Cancer, Gemini and Orion. It is not recorded who bought this unique work of art. And the rumor that the British Museum in London is its present possessor was not confirmed there.

In addition, thin and thinnest watch movements were naturally developed too: in 1909 the masters introduced two calibers which, including their bowls, measured only 7 mm (Extraplate) and 4 mm (Ultraplate) in thickness. Repeating watches were another favorite of the firm's proprietors. In 1910, 231 examples were finished. More than half—117 of them—were equipped with minute repeat striking, 110 of them with the 81-ligne caliber.

The business flourished: in 1918, the last year the two founders managed it together (Jules Audemars died in 1918, Edward Piguet a year later), 1590 watches were finished in Le Brassus. The manufacturers J. Aubert, Le Coultre and L. E. Piguet meanwhile supplied the raw movements.

Extra-flat model in current production, with hand winding (height of the movement: 1.64 mm).

After the founders' death the company, following the trends of the times, built more and more wristwatches, to which the same standards of quality applied as for pocket watches. In 1920 "AP" astonished the world of watches with one of the smallest watch movements ever built with minute repeat. The diameter of the movement was barely 16 mm, and remains at the top of its class even among today's highest achievements in micro-mechanics.

At about the same time, a woman's wristwatch was made, likewise equipped with minute repeat (movement diameter barely 18 mm). Case and band were thickly set with brilliant-cut diamonds. The striking was activated by light pressure on a diamond under the numeral 6. A real masterpiece!

Rectangular or barrel-shaped cases were especially popular among men, ultra-small watches among women. The business flourished in the Twenties as Audemars Piguet kept trying to build special complexities into wristwatches. There were rectangular wristwatches with simple calendars, center date and moon-phase indication as well as simple digital calendar and moon-phase indication. From the Swiss Jura came the rectangular wristwatch with "jump" digital hour indication and small second, as well as the woman's "simple" sport watch.

In 1927 the firm produced a wristwatch with digital hour and minute indication, and one year later two more new items (a wristwatch and a pocket watch) were shown in the catalogs: a 2.75-by 7.75-ligne baguette move-

Men's wristwatches from the Sixties.

Man's model with day, date, moon-phase indication and automatic winding; present production.

Wristwatch with automatic winding, chronograph with 30-minute and 12-hour counters and date indication; dial with subsidiary tachometer scale; put on the market in 1986.

"Quantième Perpétual Rivière" model with automatic winding, perpetual calendar and moon-phase indication. The movement is skeletal, case and band are set with 788 brilliant-cut diamonds (11.65 karats); 1986-87.

Wristwatch put on the market in 1978, with automatic winding, perpetual calendar and moon-phase indication, which has become a classic on the wristwatch market.

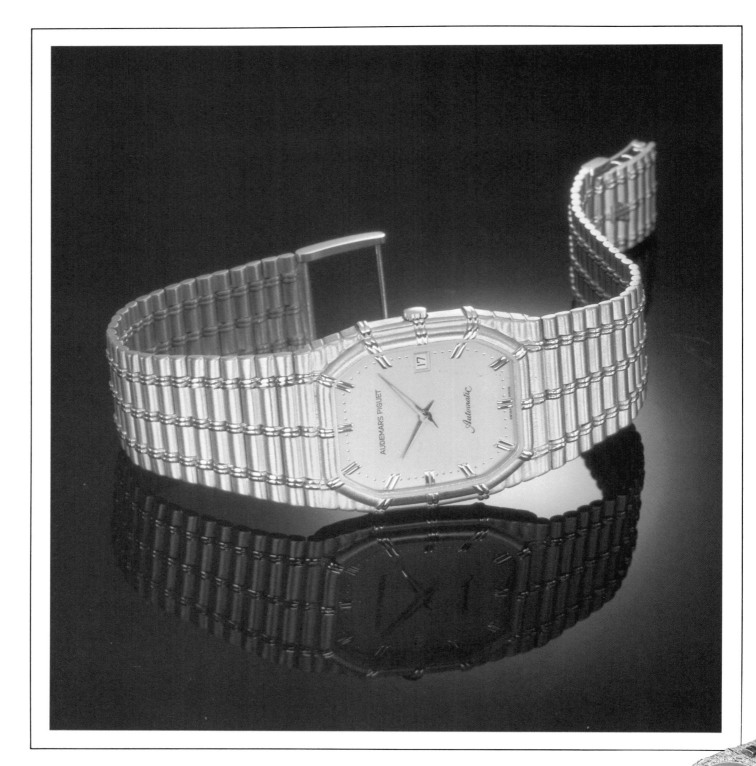

"Bamboo" model for men, with automatic winding and date indication.
"Royal Oak Rivière" for women, with quartz movement, set with 528 brilliant-cut diamonds (5.44 karats) and 48 emeralds baguettes (8.50 karats).

Man's model of the "Philosophique", a wristwatch with only an hour hand; the minutes must be estimated by interpolation. The bezel, equipped with a gold knob, can be turned along with dial and movement in order to convert to a second time zone, while the time for the original location is retained; present production.

Woman's model of the "Philosophique"; the bezel is set with fifty brilliant-cut diamonds (0.54 karat).

24 AUDEMARS PIGUET

This extra-flat wristwatch with automatic winding and tourbillon with lever escapement came on the market in 1986.

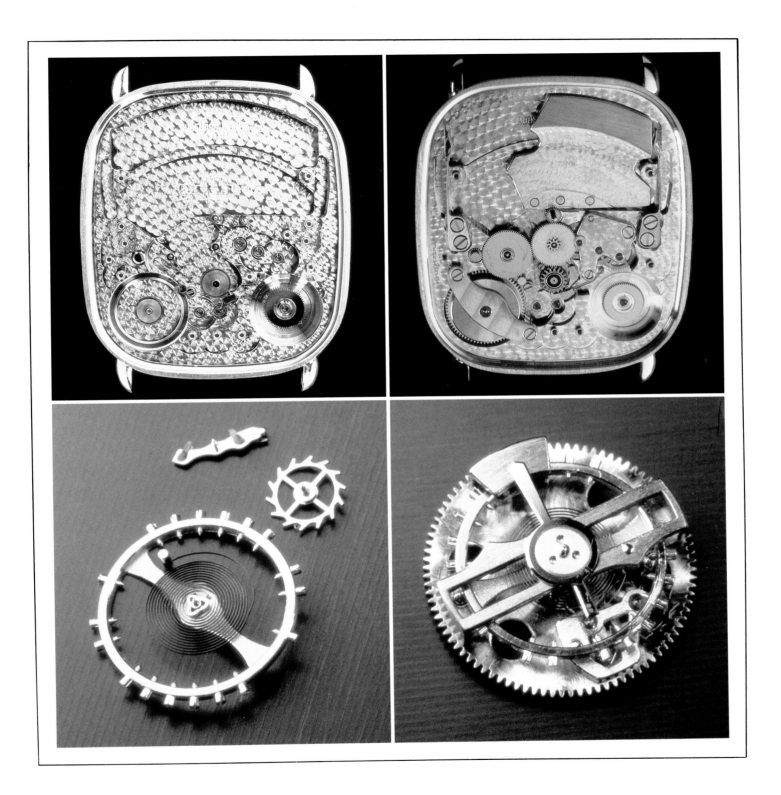

Upper left: the case bottom also forms the bottom plate of the movement.
Lower left: balance ring with hairspring, escape wheel and lever wristwatch with tourbillon.
Upper right: the wheel train and winding pendulum made of a platinum-iridium alloy are mounted.
Lower right: The tourbillon frame mounted with the escapement.

Prototype of a wristwatch first introduced in 1987, with automatic winding, minute repeat, perpetual calendar, moon-phase indication and chronograph with 30-minute register.

ment measuring only 3.6 mm high, and a movement that fit into a hollowed-out twenty-dollar piece. On the upper rim of the coin was a small button that opened the "lid", when the time was to be read. Moreover, when the coin was closed, one could not tell by the weight that it contained a complete watch.

All over the world, the finest retailers were among the customers of Audemars Piguet, including:
—Dent in London,
—Tiffany in New York and Paris,
—Cartier in Paris,
—Breguet in Paris,
—Bulgari in Rome,
—Frodsham in London, and
—Dürrstein in Glashütte and Dresden.

So the production and sales figures of 1920 (1993 watches) developed very well, until in 1928 the world economic crisis had its effect and only 769 watches could be sold. The factory had twenty-five employees on October 24, 1929, when the New York Stock Exchange crash set off the depression (737 watches were sold that year). Most countries tried to help their own economy by massive protective measures. In almost all countries outside of Switzerland the market was closed—only very few people could afford expensive watches. Old customers and retailers disappeared. People had other problems. Layoffs were announced—Audemars Piguet faced hard times. In 1930 they were still able to produce 450 watches, but then business went downhill drastically. In 1931, 54 watches were produced by a reduced work force, and the firm hit bottom in 1932: exactly two watches were made in the Vallée du Joux workshops.

The question was: how were they to go on? The heirs of the firm's founders showed that the blood of the country's staunch farmers still flowed in their veins. After a every long winter that made the isolated valley even lonelier, didn't the springtime sun appear again on the horizon? Didn't things always go on in the cyclical course of life? Thus they decided that it must go on: the tools and workshops were there, the technical know-how too, and—it really could not get any worse. The company struggled to survive. And the firm survived the worst crisis since its founding. As early as 1933, 39 watches were produced by a small staff, 33 in 1934, and by 1935 production was back on its old course: 117 watches were made in Le Brassus. In 1937 the work force numbered eleven employees, barely more than in the year the firm was founded; in 1940 there were seventeen watchmakers producing about 200 watches a year. Business climbed slowly upward again. Above all, it was the chronographs that made the company's existence secure in the beginning and during World War II; most of them were sold in the United States.

In 1934 the proprietors

Wristwatch with extra-flat skeletal automatic movement; present production.

Only in 1986 was the "Baroque" model made public. The watch glass is a single jewel, in the case of the watch shown here an amethyst, the bezel is set with 46 brilliant-cut diamonds (0.70 karat). Quartz movement.

considered a specialty of the tradition-rich watchmaking art of the Eighteenth Century and enriched their catalog with a skeleton pocket-watch movement. Skeleton watches are simply fabulous products, which require above all a truly masterful ability and sensitivity of their creators. Here nothing can be concealed under a laboriously formed and decorated case; the movement lies practically "naked" and bare before the eyes of its beholder, revealing the complete union of technical masterwork and artistic achievement.

For months the individual parts of the movement are polished under the loupe, engraved, and polished again and again by a master's hand, until the miniature components finally find their place in the function of the whole through their assembly into a grandiose watch movement. These skeleton movements, at their highest development, proved themselves to be such good sellers in the postwar years too that they are now among the firm's regular offerings and account for about 5% of its income. In addition, the positive influence on the market brought the skeptics into agreement. One thought of the past and experimented again with ultrathin movements. The result could be seen in 1946. The nine-ligne caliber 2003, with its 1.64 mm "thickness", still ranks among the thinnest movements that have given trouble-free running results, even under punishing conditions. The watch is esthetically timeless, or as an expert expressed it: "This watch has a straightforward effect, even in the regal raiment of a gold case."

Naturally there were also wristwatches that were built according to the founders' guide-

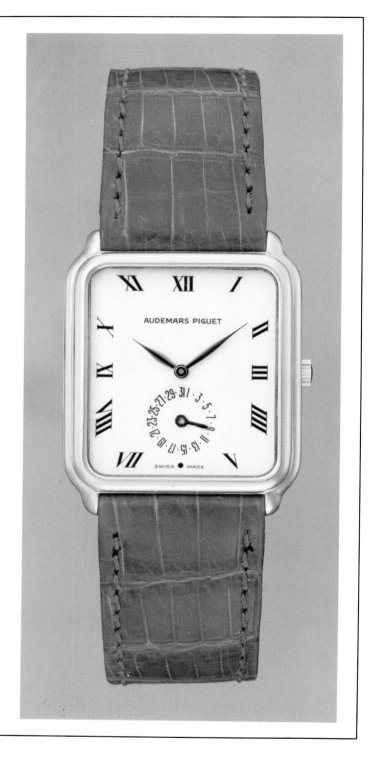

Man's wristwatch with date indication, present production.

The mechanism of the wristwatch with perpetual calendar is assembled.

lines, simply to be the best and finest, such as a man's gold wristwatch with calendar, moon phase and chronograph. And naturally the minute-repeat watches, which for so long, thanks to the tireless striving of the two founders and their successors, have been and will remain a specialty of the house.

Audemars Piguet was again on the way to the top! In 1950, with twenty-six employees, 1700 watches were produced, more than ever before in the firm's history, but still few enough to emphasize the exclusiveness of the products: six or seven per day. Then exactly ten years passed before Audemars Piguet again put a technical sensation on the market alongside their "normal" production, It was a woman's watch that made the watch world take notice in 1960. The specialists in Le Brassus created the thinnest woman's watch in the world, which was only 2.36 mm thick.

In 1967 Audemars Piguet presented what was then the thinnest watch movement with automatic winding by a central rotor. The height of this caliber 2120 was exactly 2.45 mm. The raw movement was produced exclusively for "AP" by LeCoultre, located only six kilometers away in Le Sentier, in the same valley, and became the basis for further developments in the Seventies and Eighties. Later a 1.6 mm "thick" module for an perpetual calendar was constructed, which has been offered in a particularly flat wristwatch since 1978 and has become a hit in terms of sales. For example, 550 examples of this model found buyers in 1984. But in between, in 1972, the presentation of the aforementioned "Royal Oak" took place. Steel outside, gold inside—that was the motto with which this watch won headlines in the trade press, for the rotor ring of this caliber 2120, developed in 1967 and very successful since, was made of 21-karat gold. The steel-gold and all-gold versions followed five years later, when the demand for even more exclusive models of this mechanical watch increased. It was watertight to a depth of 100 meters (guaranteed by a specially developed winding crown). Since 1983 the Royal Oak, with the same renowned qualities, has also been available as the "Day and Date", and since 1984 even with an perpetual calendar, and also—to create further refinements—decorated with selected diamonds. Approximately fifty million watches were produced in Switzerland in 1986. Audemars Piguet, with an income of merely seventy million Swiss francs from some twelve thousand watches, contributed less than 0.025% of the whole Swiss production.

Audemars Piguet employs two hundred workers today, of whom forty-four are watchmakers with diplomas from technical colleges, twenty with training at the factory, twenty-nine watch mechanics with diplomas and three jewelers with diplomas—which shows clearly that with all the firm's exclusiveness in design, the main emphasis is placed on technical perfection.

Skeletal construction of a wristwatch with automatic winding, perpetual calendar and moon-phase indication.

Classic round model in present production, with automatic winding, day and date indication. The watch shown above also offers an additional moon-phase indication.

But names and traditions are not sufficient when one wants to maintain or even increase income statistics and percentages of the market. Thus work continues on new technical developments in an attempt to exceed one's own masterly achievements again and again. But what is being worked on in the "idea kitchen" is naturally a strictly kept secret. Only when new developments have ripened into complete watches is the veil lifted. This took place in 1986, when Audemars Piguet introduced the first series-produced wristwatch with automatic winding and tourbillon. Years of research and developmental work preceded the final introduction of this watch, made of the best materials with the highest technology.

Even in the Swiss watch industry, where tradition has not been forgotten, the absence of modern technical devices would be unthinkable. Plans for the new developments of these times are not drawn only by hand on paper, but are also worked out by computers in many ways (for example, the abrasion of materials) and projected on the screen with a CAD system. Computer-guided construction methods are also used when the components are too small and "breakable" to be mounted by hand.

The smallest turning frame ever built for these wristwatches in the two-hundred-year history of the tourbillon measures 7.2 mm in diameter and is only 2.5 mm high. It is made of a material that was not yet available to the firm's founders: titanium—extremely light and stable, but also extremely hard to work. Without computer-directed electroerosion the production of this "miraculous movement" would be impossible. A modern material is also used for winding weights: a platinum-iridium alloy which gives them a massive weight. Thus automatic winding has a high degree of effectiveness, creates a winding reserve of about fifty hours, and makes a subsidiary crown winding apparatus unnecessary. So the crown can be hidden on the bottom and used only for hand-setting. The movement (32.5 x 28.5 mm) is so complicated and complex that about 150 hours are required at Audemars Piguet to finish it. An absolute high point in wristwatch manufacture was attained by the "Grande Complication" presented at the 1987 Basel Clock Fair. It possesses not only a chronograph and an perpetual calendar with moon-phase indication but also minute repeat striking. Winding is automatic, and the price is estimated at 200,000 marks.

In the last few years a new production center has been built next to the Le Brassus railroad station, a few stones' throws from the original factory: a highly modern mechanical and electronic "diversification". But even in this new realm, the guidelines of the founders are adhered to and only the highest quality is offered.

Audemars Piguet's philosophy: on the one hand, there is striving for a unique creation of aesthetic perfection and dreamlike fantasy. On the other hand, there are the requirements of form, purpose, technique

Man's wristwatch in extra-thin construction.
Classic gold man's wristwatch with automatic winding and sweep second; present production.

The classic "Royal Oak", put on the market in 1972, whose bezel resembles a porthole. With this model Audemars Piguet was able to make steel an acceptable building material for the cases of high-quality and high-priced wristwatches; automatic movement with date indication.

right—
Man's watertight wristwatch in steel-gold case.

Gold "Royal Oak" with day and date indication. Movement with automatic winding.

and precision. The symbiosis of the two creates the value of a work of watchmaking art. And from this awareness there come watches that fulfill the criteria of timelessness:

"The technical developments and the designs of today are the basis for the classic watches of tomorrow." A sentence that Jules Audemars and Edward Piguet surely could have subscribed to in 1875.

"Royal Oak" in gold with 11 brilliant-cut diamonds to mark the hours.

The technically most demanding version of the "Royal Oak" with automatic winding, perpetual calendar and moon-phase.

BAUME & MERCIER

The authentic history of Baume & Mercier did not begin—as with the other outstanding Swiss firms—in Geneva or its vicinity, but some kilometers away, near Bern.

In a small town in the Bernese Jura in 1830, just as the first great railroad line was being built in England between Liverpool and Manchester, the Baume family began to produce watches. Along with the precision customary in maufacturing at that time, the Baume watchmakers put equal emphasis on the reliability of their products. And in this way their watches achieved very quick fame in Switzerland.

In 1918—as World War I still raged in Europe—a Baume family member visited Geneva and met the almost legendary Paul Mercier. This watchmaker and jeweler had won a splendid reputation in the watch metropolis through his extravagant artistic taste and outstanding handworking ability. Beautiful and solid work linked the two masters of their art, who were so much in agreement that they decided to stride into the future together. In the very same year the firm of Baume & Mercier was founded, with its headquarters in Geneva. Blending handworking skill and technique to a complete unity at a high level of quality was the basic principle of both of the firm's founders. And even today, more than 150 years after the founding of this watchmaking company, nothing has changed. No more than three years after its founding, specialists in the field knew for sure that the little factory could turn out extraordinary work. In 1921 Baume & Mercier was awarded the "Poinçon de Genève", a coveted emblem of precious metal attesting to the finest handwork, highest quality, faultless execution and reliability in watchmaking.

The headquarters of Baume & Mercier, then and now, is located in Geneva. But its market is no longer limited to Switzerland, for the watchmakers and jewelers branched out into the whole world when World War I ended. Foreign markets were opened, and today the brand is represented in seventy countries. A large part of its production was soon transferred to La Cote-aux-Fées. Atypically, there are now production facilities even in New York. The Piaget family was undoubtedly involved in the expansive development of Baume & Mercier; in 1965 they purchased almost two-thirds of the company's shares. The reasons for the purchase are as simple as they are enlightening:

— Piaget, belonging to the leading group of Swiss manufacturers, wanted to be represented in a lower price range.

Jean-François Glauser, General Manager.
Christian Piaget, Vice-President.

Document from 1921 from Baume & Mercier, according the "Poinçon de Genève" on the basis of voluntary accuracy tests at the Geneva Test Center.

— Their own products were not to be changed, and the name of Piaget was not to be devalued.
— The partial purchase of Baume & Mercier gave Piaget the chance to control the situation without taking the full risk. Piaget was at that time not yet strong enough to take over the production 100%, as they themselves well knew. Today one can add the word "unfortunately" to this.

The takeover came at a time when Baume & Mercier were producing extraordinarily beautiful mechanical watches, precise as well as elegant, with great success. Carrying on this concept, Baume & Mercier, for example, put the thinnest automatic calendar watch in the world, made possible by a planetary rotor, on the market in 1968.

In the same year the new owners also made a venture into the realm of electronics: one of the first electronic watches with tuning-fork movement was put on the market.

In 1970 Baume & Mercier turned to quartz power and made every twentieth watch without a balance. In the following years the brand proved to be one of the pioneers of new technology and gradually phased out mechanical watches. The quartz watch movements that were 3.6 mm thick in 1977 measured 2.5 mm in 1980 and as little as 1.65 mm in 1982. As the electronic movements became thinner, they also took over a greater share of production. By the end of 1983 a milestone of development had been reached: the last mechanical movement was finished. Since 1984, after lengthy research and development, the firm has made only quartz movements, which—in view of tradition—are assembled completely by hand for the sake of enduring quality. And this in turn finds expression in their one-year guarantee of running regularity (maximal plus/minus sixty seconds in a year).

Naturally, an exceptional quality control is required. This is carried out for several days wth the most modern testing instruments, although every Baume & Mercier watch nevertheless is subjected to the official Swiss quality control.

That such requirements bring results was shown by the factory in July of 1980 with a "Riviera" from their current production: The watch was mounted on the wheel of a BMW M1 before the start of the Le Mans 24-hour auto race, withstood speeds of more than 300 kph, enormous acceleration at the start and on the straightaways, the centrifugal force that pulled at the watch, hour-long cloudbursts and the extreme heat of the overworked disc brakes, and—undamaged—ran ultra-precisely after the race.

Undoubtedly, though, the factory has won the good reputation that it possesses today particularly through their superb designs, along with the precision of their movements.

Wristwatch with date indication, from the "Riviera" line, in steel and gold, with quartz movement.

36 BAUME & MERCIER

Men's gold wristwatches with day, date, month and moon phase indication, quartz movements; 1988 collection.

Woman's two-color wristwatch with central sweep hand and date indication: case watertight to 30 meters; 1988 collection.

Men's wristwatches with chronograph, 30-minute and 12-hour indicators, as well as date indication, movements with automatic winding, cases in gold or two-colored steel, watertight to 30 meters; 1988 collection.

The basic principle of the firm's proprietors is making the highest achievements of the engineers more meaningful through aesthetic fulfillment. Thus the quartz movements are set in golden cases, surrounded by jewels, attached to exclusive bands—finished, in any case, by good old handwork, then and now, by the firm's approximately sixty employees. Faultless work, as they say in Geneva, simply cannot be done by machine. Thus the master's eyes, hands, experience and ability, play the key role in the production of every single Baume & Mercier watch. And thus the individuality of every single piece is born—an individuality that Baume & Mercier have taken to its extremity through a unique process of combination, resulting in more than eight hundred different models: different dials, different cases, and over a hundred different band designs can be combined to suit the customer's wishes. Thus it is no problem to obtain one's "very own" watch.

Another center of emphasis at Baume & Mercier is obviously the design and production of bands. The jewelry workshops have always proved to be pathfinders to new creation in cord, foxtail, link and chain straps that—if necessary—can be tailored individually to the right length at the factory according to the measurements taken by the retailer. The safety catches are constructed in such a way that unwanted opening is really impossible.

The case of this wristwatch from the "Avant-Garde" line is made of gold and tungsten carbide; quartz movement with date indication.

This wristwatch with pulse scale is intended especially for doctors; quartz movement with date indication.

Gold-plated wristwatch from the "Linea" line; quartz movement.

The "paved" dials are another specialty: massive slab plates that are set with selected diamonds by hand—several hundred for some models. Thus are the "paving-stone" dials made.

It is taken for granted by the Geneva watch manufacturers that all watches be equipped with scratch-resistant sapphire glass, and that the models marked on the bottom with a stylized fish must be watertight to a depth of 30 meters. Whether of eighteen-karat gold, steel and gold, or steel, whether dress or sport watches, the watches are made in roughly equal portions for men and women. The prices of the models in production (100,000 watches are sold in a good year) begin at about 900 Swiss francs and go up to just over 40,000 francs.

Most Baume & Mercier watches are sold in the USA. There the company has had the brand registered in the 1260 American customs offices in importing harbors to prevent unauthorized importation of the products and protect the market against counterfeit market importing. Europe, and finally the Middle East, follow in sales statistics.

Despite all the demands of fashion, the styles of the times and the trend toward uniformity, an individual Baume & Mercier style has developed, falling into two different lines: the classic and the sporting, which live up to the same standards of style and elegance. Typical sporting lines are the "Riviera", already almost a classic, and the "Jubile". Both watches are watertight, have an ultra-thin quartz movement with calendar and large second hand, and are set in steel-gold or steel cases.

One of the newest creations is sporting and elegant at the same time: the "Avant-Garde", in which the factory mounted movements in an extremely hard material for the first time: gold with tungsten carbide. Thus the case and strap are now just as scratch-resistant as the sapphire glass over the dial. Here too we find an extra-thin quartz movement with calendar and a case that is watertight to a depth of thirty meters.

Naturally there are also the timelessly elegant classics in rich abundance, chronographs in the style of the Thirties, in which the dial practically becomes an armature panel: moon-phase and date indication, centrally mounted chronograph hand with 30-minute and 12-hour register dials, small second and tachometer scale. The additional buttons control starting, stopping and zero-setting of the chronograph.

With the "Haute Joaillerie" array, Baume & Mercier offer sparkling symphonies of the jeweler's art that make ultra-precise timekeeping one of the nicest sidelines, and that cannot deny their relationship to the mother house of Piaget.

This man's wristwatch from the current production is based on classic models; it includes chronograph, 30-minute and 12-hour registers as well as a tachometer scale.

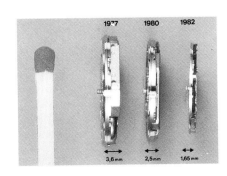

The steps of development on the way to ever-thinner watches are made clear in this picture.

BLANCPAIN

In the summer months of the early Eignteenth Century, the narrow mountain valley amid the mountains of Neuchâtel that tower over a thousand feet high offered a picturesque view. The visitor, who had come on horseback, gazed across little brooks with green-grown banks at the town of Villeret, at the approach to the mountain chain of Chasseral and the French Alps. A picturesque village, in which the mountain farmers struggled to grow grain and potatoes in the poor soil through perseverance and endless work.

When the first snowflakes floated down from the heavens early in the autumn, the picture changed. Covered with glittering whiteness, the valley sank into an isolation that lasted five to six months.

Emile Blancpain

Where modern snowplows open the winter sports fans' way to the ski lifts today, there were not even roads then. The Jura valley and its people were cut off from the outside world by masses of snow a meter high—isolated in slowly passing time. Only the light ribbons of smoke that one could see rising from the houses on clear winter days showed that life went on within the protection of thick stone walls.

But what kind of a life was it? It was limited to waiting. Waiting until the sun's rays became warmer again and made the snow on the mountainsides and in the village melt. Then the cycle of life began anew in this part of the Swiss Jura— ruled by the eternal change of the seasons.

At that time a flourishing watchmaking trade had already developed in Geneva, several days' ride away. This development had, in fact, been inspired by the reformer Calvin, who had banned the production of sacred objects by the goldsmiths, as well as by the Sun-King Louis XIV, who by revoking the Edict of Nantes in 1685 had driven the Huguenots, who dominated watchmaking in those days, into Protestant Geneva. Some of them, to be sure, settled down in La Neuveville on the Lake of Biel, where a vigorous trade quickly developed beyond the reach of the strict Geneva guild. And one day the interests of the watchmakers on the shores of the Lake of Biel meshed with those of the mountain farmers in Villeret.

The goldsmiths looked for hardworking villagers who would make parts for their watch movements as cheaply as possible, for this laborious and petty work brought them the least profit. The most was earned by the "etablisseurs", those handworkers who assembled watch movements, finished watches and sold them. The mountain farmers in turn were thankful for any occupation during the winter months, when they could seldom set foot outside the door. So it was that representatives from the city went into the valley, bringing raw materials and tools. In the spring they came riding back to pick up the finished movement parts.

One of those mountain farmers was Jehan-Jacques Blancpain, born in 1693, scion of an old established family that can be traced back to the year 1577, when the name was still spelled Blanpan. With patience and handworking ability he quickly won a good reputation with his employers, and was entrusted with more and more complex jobs. Along with that he sought an additional source of income: he traded movement parts to his employers for finished watches and sold them, becoming a "watch dealer".

Then came his lucky day: Jehan-Jacques Blancpain dismantled one of the swapped watches, understood the mechanism, made—illicitly, one might say—all the parts for a complete watch, put them together and had a watch of his own. The next followed and was sold, as was fitting for a watch dealer, which he still was "on the side". In the next winter Jehan-Jacques Blancpain built six dozen watches, saddled his horse when the snow began to melt, and rode to the small castles in the area to sell his timepieces. The mountain farmer was thus far ahead of his colleagues. They limited themselves as before, though of course with the highest perfection, to the manufacture of wheels, pinions and other individual parts, and earned an income during the winter months that was far above what they made from their laborious field work during the short summer. Prosperity slowly increased in the valley, for the money they earned could be spent on other handmade products in the village. This prosperity amazed even a Prussian officer who visited the Bishopric of Basel in 1740. He was welcomed on his travels not by "poor careworn potato-farmers, but by proud burghers in the splendid clothes of worthy noblemen and princesses"—as one of his fellow officers wrote in astonishment.

The first watch signed by Blancpain is documented in 1735. But it can be assumed that watches had already been sold under this name for ten years. A historian wrote in 1725 that the population of Villeret consisted of "350 souls (men, women and children)", and he laboriously described the occupational groups along with six saddlers, four nail-makers, four millers, two tailors and one smith, there were also two watchmakers. From this document, which is now preserved at the museum in Basel, it can be concluded that one of the mentioned watchmakers must have been that same Jehan-Jacques Blancpain who can be regarded as the founder of the world's oldest brand of watches.

In 1815 the change to a small watch factory, bearing the name of Blancpain and functioning on an industrial basis, took place under Frédéric Louis Blancpain. In later advertisements the year of 1815 is always cited as the date of the foundation of the firm, whose name was to change several more times in the following decades:
—1830, to "Emile Blancpain",
—1857, to "E. Blancpain et fils",

The first industrially produced (since 1929) wristwatch with automatic winding, named after its inventor, John Harwood, produced by Blancpain for the French market.

A man's wristwatch, the "Rolls", with automatic winding in which the movement moves back and forth in the case, made by Blancpain in the early Thirties.

—1889, to "E. Blancpain Fils", and
—1928, to "Blancpain, Fabrique d'Horlogerie à Villeret".

Just as varied as the firm's different names were the trade marks that were registered with the Eidgenössische Amt für geistiges Eigentum in Bern during the course of the years. One may well start with the assumption that only a few of the watches made by the House of Blancpain were actually signed with this name. Often a manufacturer's signature was completely missing, since the watches bore the name of a concessionaire.

After the death of Frédéric-Emile Blancpain in 1932, the era of the Blancpain dynsty in the firm's history came to an end. Since he had no male heirs, his closest collaborator, Madame Fichter, carried on the business for almost forty years under the name "Rayville SA. succ. de Blancpain"; "Rayville" was a phonetic anagram of the name "Villeret". From 1959 on the firm has been called "Manufacture d'horlogerie Rayville SA, Montres Blancpain" and has produced nothing but wristwatches, some with its own, some with purchased calibers.

Blancpain already had a considerable effect on the history of the wristwatch in the Twenties and Thirties of our century. In 1926 a prototype of the famous "Harwood", the first series wristwatch with automatic winding, was produced in Villeret, as well as the same movement in a small series for the French market.

From 1930 on the movement for the "Rolls" was built, an automatic wristwatch in which winding was accomplished by the "rolling back and forth" of the movement in the case. Such masterpieces carried Blancpain's fame throughout the world. Likewise the diver's watch which as early as 1953 was watertight to a depth of 200 meters, the "Fifty Fathoms".

In 1970 the company passed into the possession of Holding SSIH (now SMH), joining the Omega and Tissot brands.

For twelve years the brand was not activated. Watches with the Blancpain signature existed only as antique collectors' pieces, until in 1982 Jean-Claude Biver, then director of Omega, attained the high point of his life: he bought the brand name from his employers and made himself independent on January 9, 1983.

Today the young entrepreneur, who builds watches as fanatically as he takes part in marathon runs all over the world on Sundays, can only rejoice at this purchase. When he signed the contract, his prospects were not so rosy. After signing the contract, the buyer, Jean-Claude Biver, had nothing but the brand name, the warehouse and the staff. Orders were not to be seen. The original factory in Villeret did not belong to him. Watches were produced there for Holding. Biver had to look quickly for a way to keep his company and staff in business. But he had thought ahead. The "outsider of the branch" (as Biver knowingly described Biver) shared the ownership of the world's oldest brand of watch from the beginning with Jacques Piguet. The latter, in turn, is a descendant of Louis Elisée Piguet, a well-known Swiss designer, to whom other outstanding firms turned for contract work. This created instant confidence in the branch

"Fifty Fathoms" model, a diver's wristwatch watertight to 200 meters, with automatic winding, first produced in 1953. The indication by the 6 shows whether moisture has entered the interior of the case.

Extra-thin wristwatch with automatic winding, sweep second and date indication, watertight to thirty meters, available in men's and women's models.

and gave the energetic Biver another advantage: Blancpain moved to the tradition-rich town of Le Brassus—to the farmhouse where Louis Elisée Piguet had built his first watch movement. In new quarters, Blancpain found its way back to tradition. And tradition was exactly what Biver wanted to sell with Blancpain watches.

The clever and well-spoken young businessman astounded the branch with one stroke when he got results from two significant bits of experience: the trade in diamond-studded, ticking gold bars (for sale to Arab oilsheiks) was fading in 1982. The auction business in antique watches was approach-ing a high point. Thus Biver aimed at a gap in the market with marketing that stressed luxury and advertising that drew on tradition. He proceeded from the fact that old moon-phase watches often sold at auction for more than twice their already-high estimated values. His conclusion: he wanted to build such watches.

One look at their dials should suffice to recognize them as special instruments with costly mechanisms inside. That was exactly what he believed was in demand, but scarcely anyone believed Blancpain could attain such a renaissance three decades after its last technical masterworks had been achieved. Biver, who had become well acquainted with the characteristics of the branch as sales manager of Audemars Piguet after his marketing studies, revealed inspiration and conviction. His sales success soon silenced the last skeptics. The old watch brand of Blancpain was rising and attracting the attention of watch fans and Swiss luxury manufacturers in

Woman's wristwatch with complete simple calendar and moon-phase indication. Made at first with hand winding, it has also been produced since 1984 with an automatic movement.

Man's wristwatch with automatic winding, perpetual calendar and moon-phase indication, first offered for sale in 1984.

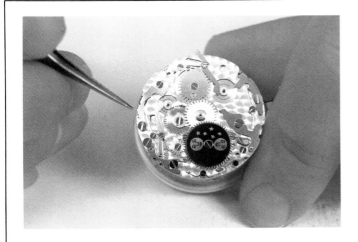

one burst after another.

The competition did not want to leave the harvest of the moon-phase field entirely to the "newcomer with tradition". But the factory in Le Brassus had already proved to be the leader in this profitable realm: the success of his moon-phase wristwatches with steel or steel-gold cases led Blancpain to further developments in the direction of complex wristwatches for women.

In 1984 the first mechanical moon-phase/calendar watch for women emerged from the farmhouse in the Vallée des Joux (hand-wound movement of eight and three-quarter ligne). A year later there was another premiere: the first moon-phase/calendar watch with automatic winding for women, in 1986 a woman's watch with the thinnest automatic movement, date and sweep second. The movement has a diameter of 21 mm and is 3.2 mm thick.

The resurrection of Blancpain today cannot be explained merely by the new owner's "courage in filling holes in the market" with his moon-phase watches.

Since the start, a variety of models from the factory in Le Brassus have become very popular among customers.

With his decision to revive technical masterworks of the past, a new philosophy for the firm was created, which Jean-Claude Biver formulates sensitively thus: "We do not repeat yesterday's history. We create tomorrow's history today." The results of this have gained noteworthy recognition among the friends of Blancpain.

There will never be quartz watches of this brand. Ornamental watches of the Fifties are just as unlikely to experience a rebirth here as the extravagant "Rolls" watch or out-of-favor cufflink watches. Round watch movements as mechanical masterpieces, always in the same round cases, this is the direction in which Blancpain wants to fill the traditions of the watchmaker's art with life now and in years to come. Thoughts of playful designs are not allowed to exist among the specialists in Le Brassus, for—as Biver sees it—variations from the round form turn a wristwatch into a fashion accessory. And that is exactly what he does not want.

This look back to tradition naturally has consequences in terms of the sizes of the wristwatches (there are only one women's and one men's size), their appearance (round, with Roman numerals on a traditionally white dial), and their materials: steel, steel-gold, gold and platinum. The real art is in the mechanical movements, that should be regarded as six different masterpieces. First we see the wristwatches with the ultra thin movements, second the moon-phase watches with calendars,

Under-the-dial view of a wristwatch with simple calendar and moon-phase indication.

Completely newly developed gold man's wristwatch with automatic winding, chronograph-rattrapante, 30-minute and 12-hour indicators as well as date indication. The movement is only 6.75 mm thick and has a diameter of barely 27 mm.

then moon-phase watches with perpetual calendars. They are followed by chronographs (but they have to be chronographs with sweep hands), the tourbillon wristwatch known as "Whirlwind" (with swinging and escapement built in a turning frame around whose arbor they constantly turn), and finally the minute repeating watch.

This same minute repeating watch was the second sensation from the factory. Two and a half years of developmental work and a million Swiss francs were invested in the first eight examples that were displayed at the 1986 Basel Exposition. The unique models excited the visitors so much that 132 of these watches were ordered directly at the firm's booth. But this brought about difficulties, if only individual pieces were to be built: only fifteen repeating watches can be built in a year by six specialists. Thus many interested people had to wait several years for one of these watches. They are assembled from more than 300 individual parts and strike the hours, quarters and minutes when a lever is pushed.

Blancpain proves that the exterior of a watch tells nothing of its value: the lever that makes the watch strike was hidden behind the bezel opposite the crown. Only the modest inscription "Répétition minutes" under the brand name on the dial hints at the special features in the technology and value of this watch.

The complicated movement is only 3.2 mm high and 20.3 mm in diameter. Every single part is finished by hand. But even with a loupe one can scarcely recognize the smallest part of the movement. It has a diameter of three hundredths of a millimeter and is only four hundredths of a millimeter long—about half as "thick" as a human hair.

The firm had to develop its own tools and build special holders to allow the master to do his handiwork. And the computer was used to calculate the ideal forms for the individual parts, the friction losses and tolerances. A Blancpain, after all, is meant to be inherited.

Handworking skill has been blended with a feeling for sound and acoustics. The time is struck by two tiny hammers that strike on two tone springs. Tuning the striking mechanism calls for the greatest precision. The hammer must strike as hard as possible but can only touch the tone spring briefly, only a few hundredths of a second, otherwise the vibrations of the tone spring will be dampened too much. In a pocket watch the sound is transmitted to the case as in a string instrument, thus amplifying the sound. But the case of a wristwatch cannot swing freely. It lies against one's wrist, for one thing, and for another, it is pulled to the sides by the strap and thus dampened. The fine vibrations inside the watch must be directed

This thin wristwatch with minute repeat came on the market in 1986. The newly developed movement has a diameter of 20.3 mm, a height of 3.2 mm, and consists of some 300 individual parts.

toward the dial, which is primarily responsible for amplifying the sound. In order to overcome these difficulties, Blacpain experimented for a long time. They succeeded, as can be seen by the subsequent production of women's repeating watches in Le Brassus.

The production of a Blancpain watch is personified: every master watchmaker works on one watch himself from the first part to the last. Thus there is no quality control by a "superior" master. If a part does not function perfectly, the master who built the watch puts his own work back under the loupe. If one asks what work the watchmaker actually does personally, then the limits disappear, for in the movement alone there are three different possibilities. Either the part-owner Piguet supplies only the raw movement from its

A 1987 model: man's wristwatch with automatic winding, perpetual calendar, moon-phase indication and minute repeat. The control lever for the striking mechanism is concealed behind the 9.

Women's and men's models of a gold-band watch with complete simple calendar and moon-phase indication. The man's watch is set with 1012 diamonds (13.06 karats), the woman's with 704 (7.39 karats).

factory next door (the Blancpain master must then prepare, polish, decorate and mount the individual parts), or decorated individual parts or complete movements are brought through the massive steel door that separates the two factories. But the machines and the movements are developed mutually, the final production takes place exclusively in the rooms reserved for Blancpain. There the watches take on their personalities when their reference numbers are engraved. This movement number, the date of finishing and the name of the master are recorded, a numerical system that was introduced only when Jean-Claude Biver took over the firm. Behind it is a system that gives the buyer information about his property. In each case the first watch has a number engraved that begins with an Arabic number 1. There are, for example, six such examples of the moon-phase watch, for it is made in steel, steel-gold and gold, each in two sizes. The repeating watch is an exception, for the first 25 examples were to be made especially attractive: these models were given Roman numerals from I/XXV to XXV/XXV for the firm's 250th anniversary. Only then was the Arabic 1 used.

At most 2700 watches are delivered yearly. A significant increase in production is not planned, for the often-cited tradition speaks against it. Jean-Claude Biver directs the firm as once the first Blancpain did, who planned, built and sold his watches himself, as a patriarch, although the sixteen master watchmakers in Le Brassus have a free choice of their work time. They can build "their" watch whenever they want—by day, by night, or on Sunday. To make this possible, each one has his own key to the factory, but Jean-Claude Biver has sworn he will not give out more than twenty-five keys. Another feature of his business practice is that the chief knows "his" 150 worldwide sellers by name and invites each one of them to dinner every year and a half. This means that the "new Blancpain" is traveling around the world nine months of every year, although the main markets for his watches are Switzerland, Germany and England. The USA is next to be "conquered", while the Middle and Far East play less significant roles. Biver puts much faith in his own decisions. He himself decides who gets the next finished watch, for production is done exclusively to order. He makes the decision to introduce a new advertising concept when he is no longer pleased with the old one. And Jean-Claude Biver, the reincarnated Blancpain, has defined his future goals clearly: first the chronograph with sweep hand, then the wristwatch with tourbillon. And then, when the six masterpieces are fully developed?

"That's the end", says Jean-Claude Biver, meaning: then the various movements of the six masterpieces will be combined to make new masterpieces...

The nine-ligne watch movement with hand winding and minute repeat.

The farmhouse in Le Brassus, headquarters of Blancpain S. A.

BREGUET

During the Battle of Waterloo, the field marshals had problems with time. One waited for Grouchy, who did not arrive, another for Blücher, who decided the battle. Napoleon and Wellington kept looking nervously at their watches—the only thing that united them was that each relied on a Breguet. Ironically, the victor, Wellington, bought the pocket watch originally ordered by Bonaparte—engraved with a map of Europe.

The two generals were not the only great ones who trusted in these watches that were as precise as they were elegant. Watches engraved with the "Breguet" name ticked for the Lords of Chesterfield, Londonderry and Beauchamps, the Dukes of Marlborough, York, Polignac and Praslin, the Kings of Bavaria, Naples, Spain, Holland, Westphalia and Tuscany, as well as for Louis XVI, tsars and empresses. Alexander von Humboldt took them on his voyages of discovery. Sir Winston Churchill relied on his "Montre Chronograph" (No. 1886), just as Queen Elizabeth II now relies on her "Sympathique" (No. 666). Breguet watches found their way into Nineteenth-Century literature in the works of Balzac, Stendahl and Dumas, the last of whom equipped the Count of Monte Cristo with one. In Jules Verne's "Around the World in Eighty Days", a Breguet took care of timing for Phineas Fogg.

The man who is said to have completed two hundred years of watchmaking art in fifty was born Abraham-Louis Breguet at Neuchatel, Switzerland, in 1747. His father, who had fled from Paris at the time of the "Dragonades", died when the son was just ten years old. The mother married again, to a watchmaker from Neuchâtel who, like her, was a Protestant of French descent.

Abraham-Louis showed an interest in the watchmaker's craft, and had—as his stepfather estimated—an extraordinary talent that could develop only in an elevated atmosphere. To his parents, that naturally meant France.

In Versailles, near the French court, the boy of fifteen began his apprenticeship. Just after he finished it, his stepfather and mother died, one shortly after the other. Abraham-Louis Breguet was left on his own. He worked as a watchmaker, built, among others, a watch later worn by Marie-Antoinette, and studied mathematics as well, because he had recognized that this science was the basis of watchmaking at the highest level.

In 1775 Breguet married a prosperous bourgeois' daughter. Her dowry made it possible for him to open his own workshop. It soon proved that the connections he had made with scholarly people during his studies were

Abraham-Louis Breguet.

worth money: his former professor had him admitted to court, Queen Marie-Antoinette was fascinated by the unique watches that wound themselves, and spoke so well of him that Louis XVI bought several watches. He gave one of them to the mariner Bougainville, who was just organizing his great expedition to the North Pole. Two essential requirements for the further development of the workshop had been achieved: Breguet had found access to the powerful and rich, and his inventive talents showed him to be a technical genius. His greatest ideas were:

— *la montre perpétuelle,* the pocket watch with the automatic winding. It is based on a development of A. L. Perrelet which Breguet perfected so well that it could already be produced in series in 1787.

Gold enamel pocket watch with automat.
Gold pocket watch with eccentrically designed skeleton movement; model from present production.

Gold enamel pocket watch with touch hand, "Montre à tact". The pearls on the case rim are the hour markers.

52 BREGUET

Old pocket watch with date, moon phase and winding indications.

BREGUET 53

Man's wristwatch with automatic winding, moon-phase and winding indication. The dial and arrangement of the indications are designed like that of the antique model.

— *la répétition*, the pocket watch with striking on demand. This was necessary to tell the time in the dark. Breguet refined the invention of D. Quare, achieving absolute eliability through improvement and simplification.

— *la montre à tact*, the watch with the touch hand, a circling hour hand that can be felt on the outside of the case along with hour knobs, making it suitable as a watch for the blind.

— *le tourbillon*, the turning-frame watch that achieved an amazing degree of precision. Breguet began with the theory that the gravity of a pocket watch that was almost always carried vertically led to running deviations. He wanted to rule out all differences of position with the tourbillon. He developed a small "clock within the clock", meaning that the balance and escapement turned on a common axle within the movement, for example, once a minute. This eliminated most deviations caused by differences in position.

— *le parachute pour le balancier*, the shock resistance for the balance bearings. Before its invention, most pocket watches were easily damaged if they fell to the ground. Breguet found a way to put the bearing jewels on springs, so that the ends of the balance arbor could move sideways as well as axially but automatically return to their original position again.

— *la pendule sympathique*, a table clock with a half-moon shaped fork that held a specially constructed pocket watch. The latter was automatically wound and set to the exact time overnight by the table clock.

These inventions of Breguet laid the foundation for his career under the rule of two Bourbon kings, three governments of the First Republic, and the reign of Napoleon. No matter who was in power in Europe, he wore a Breguet. The master, who built the first watches with perpetual calendar and moon-phase indication as early as 1795, dedicated himself not only to further technical development; he gave his watches elegance and considered styles and special requests, he surrounded his works of art in machined silver and gold, he created carrying cases of Morocco leather and provided spare parts: a second dial, a spare glass and a mainspring. But he also fought against the first counterfeiters of Breguet watches. Working with the engraver Jean Pierre Droz, he invented the "secret signature" on the dial. This secret sign could be read only when one held the watch to the light and looked through a loupe. To be very safe, all watches that left the premises were registered in thick books, so that their authenticity could be proved later.

These books still exist and are carefully preserved in a Paris vault. Every watch is recorded in them with the name of the watchmaker, the cost of production, the date of sale and the name of the buyer. The name recorded

Gold-band wristwatch with bezel and band attachments set with diamonds. Ultra-thin hand-wound movement.

Man's skeleton wristwatch with mechanical movement, bezel and band attachments set with diamonds.

Man's gold wristwatch with automatic winding by a gold rotor. Eccentrically located dial, day, date and moon phase indication. Silvered machined dial in typical breguet style. The case is channeled on the side.

A wristwatch with the same dial design was made by Abraham-Louis Breguet in 1812; the watch shown here was made in 1988.

most often on the first pages: Marie-Antoinette. With her preference for Breguet watches, she created the economic foundation of the factory, as can be seen clearly from the records. The queen bought six watches at a time, and the court society imitated her. An ornament that she never got to see, because it was completed only after her death, bears her name: the "Marie-Antoinette", which was ordered by a guard officer with the stipulation that all known complications be built in. The brass usually used in the movement was, whenever possible, replaced by platinum and gold. This watch later disappeared without a trace after a break-in at the L. A. Mayer Memorial Institute in Jerusalem. In fact, of the five thousand watches that were built under the aegis of the form's founder, barely a thousand are still known to exist in all the world.

Five years before its two hundredth anniversary, in 1970, the firm was bought by two Paris jewelers, the brothers Jacques and Pierre Chaumet. They decided to make the brand, which had lost some of its meaning, come to life again in the spirit of its founder, and they succeeded.

In 1987 Breguet finally passed into the possession of Investcorp. The headquarters is still in Paris. But for some time the watches have been built in Switzerland, the country in which Abraham-Louis Breguet was born. A studio was opened in Le Brassus, at which the watchmakers—partially with tools from two centuries—build splendid watches. Precision, care and principles of design have become valid here again, as once they were to the "father" of the firm.

A Breguet watch can be identified reliably by five unchangeable features:
— the edge of the cases are milled (like the edge of a coin) and the bezel and back cuvettes are snapped on,
— the hand-finishing of the metal, which Breguet himself had used, is still used today for all dials, whether they are made of platinum, gold or silver,
— The name "Breguet" and an individual number are engraved on every dial,
— the hands with the hole at the end or tip (commonly known as "Breguet" hands) are made of blued steel to original designs.

The various wristwatch models are created by traditional production methods: watches with simple or perpetual calendar. A women's model shows the position of the moon in a wreath of diamonds. An exclusive men's model has not only a perpetual calendar but also moon-phase indication and minute repeat. Skeleton watches appear in all their timeless beauty.

How does Sir David Salomons, who owned the "Marie-Antoinette" before the spectacular theft, express it? "Owning a Breguet is as if you have in your pocket the soul of a genius..."

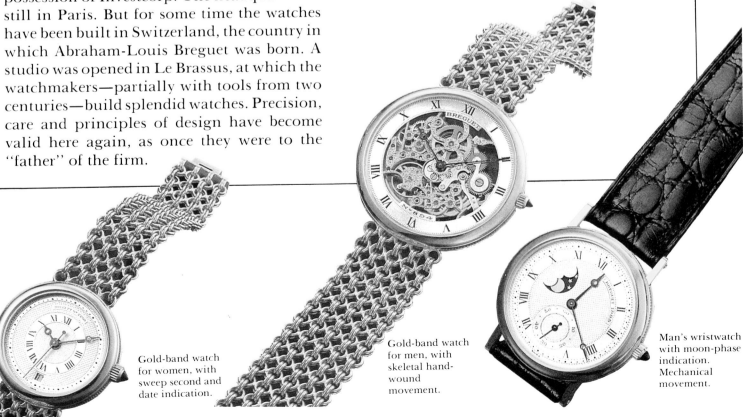

Gold-band watch for women, with sweep second and date indication.

Gold-band watch for men, with skeletal hand-wound movement.

Man's wristwatch with moon-phase indication. Mechanical movement.

56 BREGUET

Man's wristwatch with automatic winding, perpetual calendar with leap-year and moon-phase indication. At this time, about 35 watches of this model are produced yearly.

Watchmaker at a rounding up tool.

Under-the-dial views:
Upper left: wristwatch with complete simple calendar and moon-phase indication.
Upper right: wristwatch with perpetual calendar, leap-year and moon-phase indication.
Lower middle: wristwatch with winding and moon-phase indication and a simple calendar.

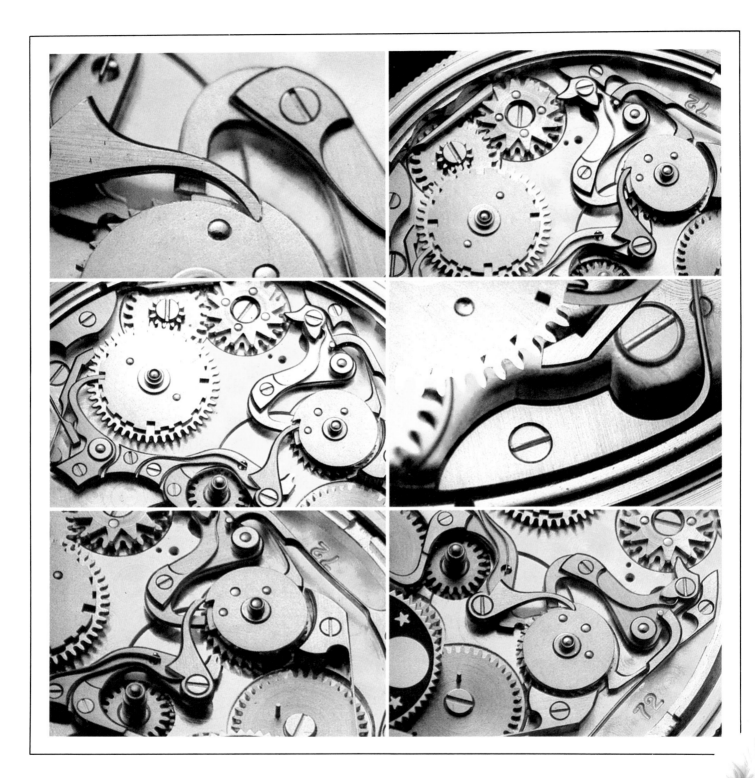

The mechanics of a perpetual calendar:

Upper left and right: at the end of the month, a ratchet switches the month pulley and thereby the date.

Center left: a point touches the step on the four-year wheel representing the February of a leap year.

Center right: a locking spring holds the four-year wheel motionless in its position.

Lower left: the switching ratchet shortly before the switching process at the end of a month.

Lower right: simultaneously with the date wheel, the month star wheel is moved one position farther at the end of the month.

A watchmaker at his workbench, busy with the finishing of a wristwatch with perpetual calendar.

BREGUET 59

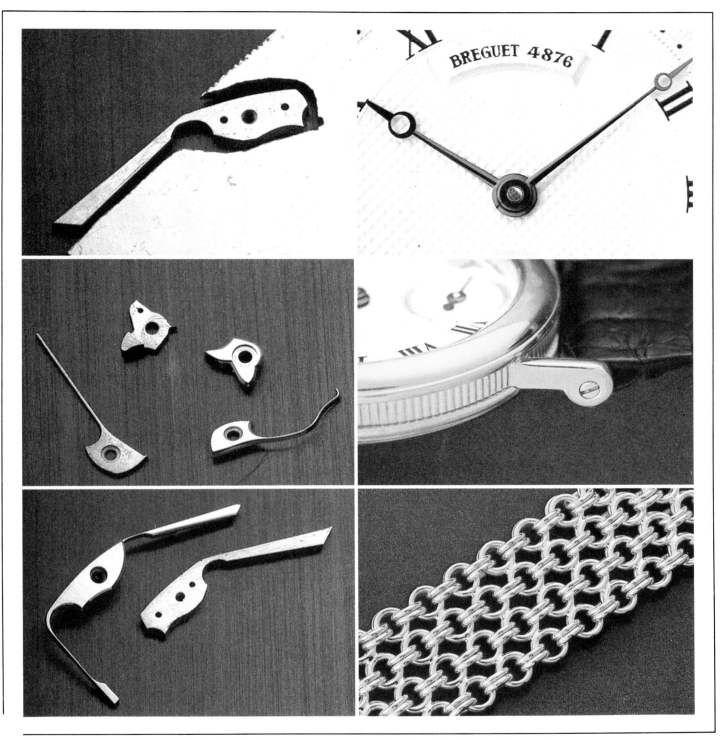

Gold Breguet clasp.

Upper left: a raw locking spring is cut out of a steel plate.
Middle and lower left: a comparison of raw and hand-finished parts.

Upper right: detailed view of the hand-finished dial and blued hands.
Middle right: the typical milling of a Breguet case.
Lower right: gold band with the so-called "Breguet links".

CARTIER

Fifty detectives and six lawyers in all the world are entrusted with just one job: to expose forgeries!—They are very successful at it. Several tons of such watches have been flattened in presses or destroyed by road rollers after more than two thousand forgery cases were won. Cartier, the most forged brand of watches in the world, has become a status symbol. Nobody thought of that, certainly, when the firm was founded in 1847.

Louis-François Cartier was born in Paris, the son of a powder horn maker. His father had learned handwork while a prisoner of war in Plymouth, England: alloying, chasing and leatherworking, before he returned to the capital of France. And naturally the son had always looked over his father's shoulder as he worked. Working with his father at first out of enthusiasm, he developed such thorough ability in the course of time that he gained an apprenticeship with the highly regarded Paris jeweler Adolf Picard. A few years later he was a jeweler's assistant and had surpassed his master in ability. In 1847 the 28-year-old Louis-François Cartier took over his teacher's studio.

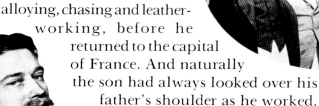

In 1851 an artist at living became the Emperor of France: Napoleon III, under whose childish rule opulent festivals were held, to which the great and rich of the world streamed. Only someone who was really prominent could count on an invitation. In 1853 Cartier, who had made a profit of 9252 francs the previous year, moved to the vicinity of the imperial palace and opened a salesroom next to his studio, in the same year in which the Emperor Napoleon III married the Spanish countess Eugenie of Téba.

Creators of fashionable clothing and jewelry hoped for the best of business on account of the wedding, for the countess was famous for her love of luxury. The city council of Paris authorized 600,000 francs for a wedding necklace with a 23-karat heart-shaped diamond, which was to be the gift of the citizens of Paris. But the Empress declined the present, and Cartier, who had promised himself noble customers above all else with his move to the vicinity of the court, saw his hopes of adding to his firm's title the words "by appointment to Their Imperial Majesties" dwindle.

Then came help from a gentle hand. Through Countess Nieuwerkerke, who had bought 55 of his pieces of jewelry within three

Louis Cartier, drawing by Friant, 1904.

Pierre, Jacques and Louis with their father Alfred Cartier.

years, Cartier made the acquaintance of her husband, who in turn was well acquainted with Princess Mathilde, a cousin of the Emperor. And so it happened that Cartier did indeed become a supplier to the court. In 1859 the Empress ordered a silver tea service from him.

In the same year, Cartier rented quarters on the Boulevard des Italiens, in what was then the "Greenwich Village of Paris". He became the darling of the rich Parisians, lived a princely life, made music in his salon in the morning, and designed jewelry that corresponded to the taste of his clientele. Cartier's jewelry was characterized by a light, airy plant-and-fruit style, very different from the English-style ornaments of the tradition-rich jewelers.

In 1859 the jeweler came into contact with watches for the first time when he bought up historic pocket watches for his prominent clients. In 1874 his son Alfred Cartier took over the business and expanded it considerably. A particular success took place four years later, when he was able to win the Duchess of Wagram as a customer. It was Alfred too who constantly expanded the firm's stock of watches and finally had a few Cartier watches made. They came from the finest firms in the business, and since 1893 mainly from Vacheron & Constantin in Geneva, who supplied whatever Cartier ordered for his customers, from a simple steel bicyclist's watch to finely decorated pocket watches. Among the prominent people who went in and out of the shop four years after the opening of the Eiffel Tower were such famous names as the Comte de Paris, the Prince of Saxe-Coburg and Pedro II, ex-Emperor of Brazil.

In 1899 the business was moved to the Rue de la Paix, the most famous luxury shopping street in the world, where at that time all the finest jewelers and most influential fashion houses were located. Cartier's offerings were now here to be admired. Business went so well that for the first time a yearly profit of more than a million francs could be aimed for.

Alfred's son Louis Cartier, whose particular interest was the old watches of the Eighteenth Century, entered the firm. His goal was the creation of the firm's own watch production. It was not hard for him to find customers, for he married Andrée-Caroline Worth, the daughter of the most famous fashion designer of the day. An exclusive list of clients fell into Cartier's lap as a wedding present, a real stroke of luck.

In 1904 Cartier made the acquaintance of the Brazilian aviator Alberto Santos-Dumont. The firm likes to describe this event as the birthday of the modern wristwatch. Inspired by complaints of the unhandiness of pocket watches in flight, Cartier designed a flat

Pocket watch, gift of King Leopold of Belgium to Alberto Santos-Dumont in 1901.

Pendulum clock with calendar and moon-phase indication, 1924.

"Santos" model, produced from 1911 on.

wristwatch for his friend Santos.

And its design was so timeless that it is still produced today in almost unchanged form. During his record flight in 1907, Santos wore this watch on his arm, equipped with a movement by Edmond Jaeger.

Edmond Jaeger, who had learned his trade from Breguet, was already working for Cartier since 1905. In 1907, the year of the record flight, Edmond Jaeger signed an exclusive contract, which was meant to begin the rise of Cartier watches. He contracted to supply watches for at least 250,000 francs per year for fifteen years, and Cartier contracted to take them. At the same time, the fashion czars of Paris had decided to swear off long sleeves. The watch on a lady's wrist thus became a particular center of attention, and the victory march of the noble Cartier watches began.

Father Alfred Cartier and his three sons Louis, Pierre and Jacques Cartier directed the firm. They took long trips. The walls of their shop in Paris were filled with diplomas that named Cartier as the supplier to famous royal houses. The branch in London that was opened in 1902 was joined in 1908 by two more in St. Petersburg and New York. The American branch moved to Fifth Avenue in 1917, when Cartier obtained the building in a very unusual way: the jewelers traded a necklace with 128 black Oriental pearls for the five-story Morton-Plant Palace!

The development of the firm continued to progress:
— 1910: Louis Cartier invented the folding clasp for wristwatches,
— 1912: the first Baignoire and Tortue models appeared,
— 1917: the Tank L. C.,
— 1930: the Ceinture model,
— 1932: the first watertight luxury watch. (The Pasha of Marrakech had complained that he never knew what time it was while swimming in his swimming pool.
— 1933: the Vendome.

A sudden break in this upward trend came in 1942 with the death of Louis Cartier. His successors made headlines only once more, when in 1969 they bought a pear-shaped diamond for $1.05 million and sold it to Richard Burton four days later at a $50,000 profit. He made a present of it to Elizabeth Taylor, making her 69.42 karats heavier.

A group led by Joseph Kanoui took over the firm in 1972. Robert Hocq, originally a vegetable dealer, then a gas cigarette-lighter millionaire, became its president. The new chief's first comment: "I have the feeling that I've married a cadaver." Robert Hocq reunited the three Cartier branches in Paris, London and New York, which had been separated through many transactions, into one firm, and installed Alain Perrin as marketing manager. This former antique dealer turned the firm completely around. He developed the "Les Must" program, which nevertheless would never have been possible without the genius

"Santos" model for women.

Diamond-set "Tortue" (turtle) model, 1913.

center
Oval model "Baignoire", whose origins go back to 1912, and which ranked after the "Tortue" and the "Santos" among the best-selling Cartier wristwatches.

Man's rectangular gold wristwatch, "Quadrant" model.

Etui watch with closing dial window.

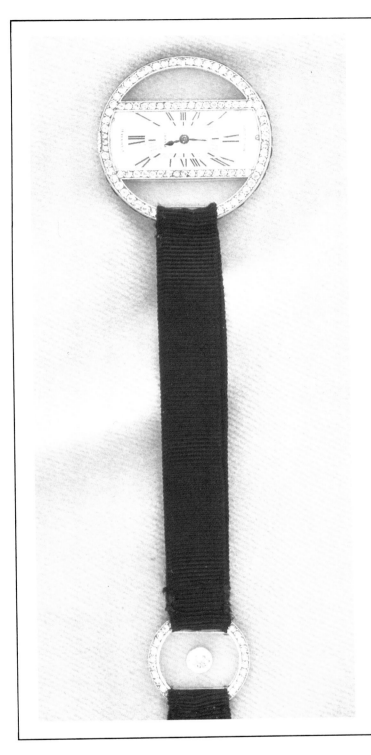

of Louis Cartier, for everything that the nearly dead Cartier giant put on the market in this line had its origins in Louis' sketchbooks.

With the rise of the business, of which Alain Perrin is now the president, the demand for Cartier watches grew. The "new" Santos developed into a bestseller, and simultaneously became the most imitated watch in the world. It was the first Cartier watch to unite gold and steel, had a sporting atmosphere, was watertight, with automatic winding, and hand-polished steel screws united its case with its watchband. The construction of today's Santos includes twenty-nine steps of assembly, in which the 99 parts of the Santos movement require 1500 separate work processes.

Cartier is the only watch manufacturer in the world who does not allow his movements

Pendant watch of platinum, decorated with diamonds, 1913.
Octagonal gold wristwatch.
Large "Baignoire" model.
Square version of the "Vendome". The design of this wristwatch with central band attachment originated in 1914.

Gold "Santos" model, for men, above; for women, below.

The "Tank" model, designed in 1917 and put on public sale in 1919, in present-day form with date and moon-phase indication.

The classic "Tank".

The stirrup-shaped "Calandre" model.

The Vendome" in white, yellow and red gold, in small and large versions. The design for this model goes back to 1933.
The "VLC" model.

The "Santos carré galbée" with date indication, in steel and gold.

The small model of the "Panthère" in gold, enameled and decorated with diamonds.

"Santos carrée" in steel and gold.

"Panthère" in gold and steel, with date indication by the 5.
"Baignoire 1920" model with gold "Casque d'or" band.
"Panthère Vendome" with sweep second and date indication, in steel and gold.

In 1933 Louis Cartier designed a watertight wristwatch for the Pasha of Marrakech; it was the forerunner of the "Pasha" model shown here, the case of which is worked from a gold bar.

CARTIER 69

"Pasha" models with complications, from left to right: Model for golf players with counters, quartz movement.
The "Pasha" with chronograph, 30-minute and 12-hour registers and date indication, quartz movement.
The most complex "Pasha" with automatic winding, perpetual calendar with leap-year indication, and minute repetition.
Model with alarm, date and moon-phase indication, quartz movement.

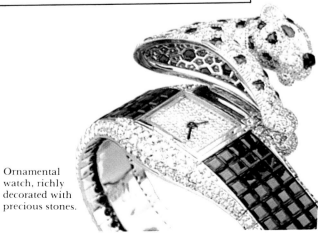

Ornamental watch, richly decorated with precious stones.

to be oiled. Patented special greases are used in Cartier's own lubricating methods, in order to reduce friction and thus improve precision. For quartz movements Cartier uses individual modules of various manufacture, in order to be able to select them for their own specific requirements. The client thereby has a special advantage: no supplier, no concessionaire, but only Cartier is responsible for the reliability of the quartz movement and guarantees its quality.

Counterfeits of currently produced Cartier watches are easy to recognize:
— Real Cartier watches are sold only by authorized watchmakers and jewelers or in Cartier boutiques, and not under the regular price,
— Cartier watches have lifetime guarantees,
— The Cartier name is hidden in the 7 or the 10 on every dial.

The ornamental watches illustrate Cartier's principle of combining luxury with necessity. The precious stones in these works of art are usually in baguette cut, to allow them to sparkle in a more slender shape. Here too, the same signs of quality apply to both case and movement. The "Tortue", "Tank" and "Vendome" versions are likewise based on designs by Louis Cartier:
— The "Tortue" (turtle) was designed after trips to the Far East, to Bali and Greece. He had found that gravestones and temples were repeatedly decorated with the figure of a turtle. At home, in the water and on the land, they were regarded as wise, strong and far-seeing. Their shell symbolized the heavens, their flat underside the mirrorlike ocean. Louis Cartier tried to express all that in his "Tortue". In 1982, exectly sixty years after the initial creation of this line, a new, extra-thin version of the "Tortue" brought the line back to life.
— The "Tank" was designed by Louis Cartier in 1917, patterned after the appearance of the army tank. He saw it as a new example

The "Chinese Tank" (1922 model) in gold, decorated with diamonds.

The "Tortue" in gold with central attachments, case and band decorated with diamonds, pearls and rubies.

The classic "Tank", band with Cartier folding clasp.

of speed, achievement and resistance. In the manufacture of this model, one feature attracts special attention: the two most important components formed by the case are numbered from the start and always worked together. At the end these "twins" form the complete, perfectly blended case.

Until a few years ago, the "Tank" was the only watch allowed to bear the "L. C." signature of Louis Cartier. Today these two initials mark the eighteen-karat gold watches.

—The "Vendome" goes back to the Thirties and a coach trip through Paris by Louis Cartier and Ernest Hemingway. On the way he noticed the shaft attachment. Cartier, who designed this model on that basis, later told that he got the idea at the Place Vendome, which gave its name to this line, which first came on the market in 1933. In 1981 the "Vendome" with quartz movement was added to the program.

—Since a few years ago, one can again wear on one's wrist what the Pasha of Marrakech once wore: the "Pasha" of eighteen-karat massive gold, with automatic winding, and watertight to a depth of 100 meters, with a golden grid to protect the glass and an unscrewable protective crown with a sapphire cabochon. In addition, the "Pasha" is one of very few watches offered with a special leather band that is insensitive to water. It is noteworthy that this watch exists only in a man's model.

Today the firm is experiencing its second high point. A yearly profit of over 300 million dollars, 120 "Les Must" boutiques all over the world, ten of its own jewelry shops, two thousand employees and seven thousand retail agents speak for themselves.

And golden years are still to follow, if one believes the Cartier president: "Cartier is No. One in its field today, in reputation, in quality, in sales and in profit. In twenty years Cartier will be ten times as great as today."

The "Pasha" for gold players, shown larger than life-size.

The gold "Baignoire" model.

CHOPARD

The Swiss federation was just twelve years old, and the world still had to wait a year for the telephone to be invented, when in 1860 the son of an old established watchmaking family went into business independently in Sonvillier. Louis Ulysse Chopard set up his own watchmaking workshop in an old half-timbered house and had the firm's name, "L.U.C.", painted on the facade.

Thus was a new brand of watch born in the Swiss Jura, for Louis Ulysse's forefathers had—like so many masters of the trade—worked only as employees until then. The ambitious young Swiss, though, strove for more precision in the smallest details. And thoroughly precise were the watches that came from Sonvillier. The Swiss railroads, that were already famous for their punctuality at that time, became L.U.C.'s most important customers around the turn of the century.

But the markets in Switzerland and the neighboring countries soon became too small for Louis Ulysse. He firmly believed that his super-exact watches ought to find customers in the Orient too.

In 1912 he packed his most beautiful pocket watches into his dark brown sample case and set off on an adventurous trip to Russia. The undertaking was a complete success, and the sample case was empty when the Swiss watchmaker came back to his native land. Chopard watches ticked at the court of Tsar Nicholas II from then on.

The business success caused Louis Ulysse to make plans for expansion. He packed his workbenches and tools into big crates and moved the firm's headquarters to Geneva in 1920. From this location Chopard anticipated international contacts without exhausting travel. For many years Geneva had been the center of the luxury watch business. The first Geneva workshop was set up in quarters on the Rue de Miléant. It was not to be Chopard's last move, but from the start the fortunes of the L.U.C. brand rose. The motto "quality through handwork" finally became the principle of the firm.

The chronicles show no further remarkable events until 1963. The company in Geneva was directed according to the maxims of its founder. Then his grandson, Paul André Chopard, had to make a difficult decision: none of his descendants showed any inclination to continue the Chopard watch firm. Should he forget the family tradition? Simply give up? Sell out? And if so, to whom? Those were plaguing questions that gave the eighty-year-old director of the firm sleepless nights. Until one day a German, of all people, wanted to secure the future of the old established Swiss brand. This German was named Karl Scheufele, and his interest in Chopard did not come out of thin air... So let us turn the clock back a few decades.

In the days when Chopard's founder, Louis Ulysse, was setting out from Switzerland to Russia, a master of his trade was also packing his suitcase in the little Black Forest village of Birkenfeld, near the goldsmithing city of Pforzheim. The jeweler Karl Scheufele went at his work just as carefully when he packed his sample case with bracelets, necklaces and rings. But his destination was in the opposite direction: Karl Scheufele headed for the New World, the United States.

At first the high society of New York did not

Front and rear views of the oldest existing pocket watch by Chopard, signed "Chopard à Sonvillier".

know who was coming. But in a few weeks the name of Scheufele was on everyone's lips. "The young guy from good old Germany", who had sold his first pieces of jewelry in a hotel room, was handed on from party to party. So his wares quickly found buyers. And the first similarities between the Chopad and Scheufele families could be seen already. For in the Black Forest too, the formula for progress was: "Handwork is the basis of success." Then as now, jewelry was made strictly by hand.

But now back to the year 1963, in which, as said, Paul André Chopard spent sleepless nights worrying about the future of his watch company. Karl Scheufele, likewise the grandson of the firm's founder, also had thoughts about the future. He wanted to round off the jewelry production of his house with valuable watches. But 'Scheufele watches" were not his goal.

It should be a brand that was as rich in tradition and as dedicated to quality handwork as his goldsmithing firm. The choice was—predictably—Chopard. Karl Scheufele and Paul André Chopard met and assured the future of the Chopard watch factory, which was taken over by the Scheufele family with all its goods and chattels, living and dead. It was not just a good deal for Paul André Chopard; he could also assure the customers of his firm that everything would continue in

Trademarks of the firm of Louis Ulysse Chopard.

Designs for ornamental wristwatches, circa 1920.

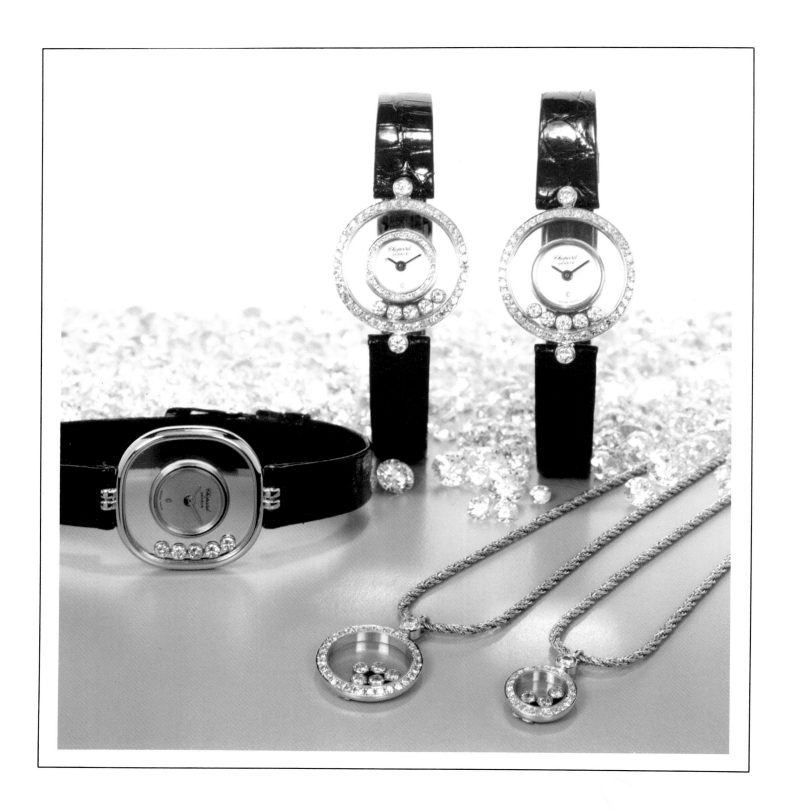

Models from the "Happy Diamonds" line.

CHOPARD 75

Man's gold wristwatch with retrograde hour indication; quartz movement.

Man's classic gold wristwatch with hand winding and small second near the six; 1988 model.

the spirit of the founder, emphasized by the fact that he continued to work in the business until his death in 1968.

The Chopard brand, which until then had found its buyers unspectacularly but successfully, was aided by the artistic inspirations from the Black Forest, the traditional striving for perfection and the high quality of its products in gaining ever-greater popularity. This strategy and new marketing brought a steep upswing for the traditional brand. New watchmakers and jewelers were hired. In 1968 Chopard moved, but relocated its studio within the city walls of Geneva. In 1974 the workshops and offices were moved again, into the firm's own building (already expanded since then) in Meyrin, on the outskirts of Geneva. One hundred fifty working positions in twenty branches were then listed in the books of the personnel department. The cooperation between Meyrin and Birkenfeld grew ever closer. The movements and golden cases were produced in the Swiss factory. The bands, jewels and ornamental accessories came, and still come, from the Black Forest. The cooperative work was successful, for Chopard expanded steadily:

In 1975 "Chopard-France" could be opened in the Rue du Faubourg Saint-Honoré in Paris, and a year later the "Chopard Watch Corporation" in New York's Rockefeller Center. An entrance to the Asiatic market was found through a representative in Hong Kong.

Now as then, the House of Chopard is dedicated to tradition. That is shown in its every move, but it does not rule out progress, as shown, for example, by the video surveillance of the firm's headquarters. Even when the cases of costly ornamental watches are cut out of massive blocks of gold or precious stone with the help of modern laser technology, the "good old days" are still present.

This can be seen in the conference room, where the wall clock, in "St. Moritz Design", shows the normal Central European Time even while summer time dominates Europe. Large photos of the products, ashtrays, notepad holders and mirrors in the style of the house decorate the offices and document that, with all the love for handwork and for the precision of the watches, the decorative effect comes to the fore. The firm's motto today:

Wristwatch from the Sixties.

A watchmaker at the Geneva studio, mounting a metal band.

"We don't build watches to measure time. We build watches for all time." Or, to quote Junior Manager Karl Scheufele: "A watch becomes a jewel with precious stones."

Twenty thousand watches (30% with mechancal, 70% with quartz movements) are mounted yearly by Chopard and sold worldwide by some 2500 dealers. Europe (Germany, France, Italy), the Middle and Far East, and the United States are—in that order—the most important markets.

The raw movements are supplied by three firms: the mechanical by Frédéric Piguet and LeCoultre, the quartz by Ebauches S.A. All raw movements are improved and—depending on the model—completed by adding the mechanism for an perpetual calendar or moon-phase indication by hand at Chopard workshops. While the dials and hands also come from suppliers, the Chopard staff is proud that all further work, to the smallest detail, is done by their own specialists. This begins in the tooling works, where the stamping forms and special tools are designed and built, and continues in the design department and the firm's own foundry.

Perfection is asked—as much as economy allows. During casting and machining the workers must put on the firm's coveralls, which are removed at day's end and cleaned in special machines once a week. The reward for Chopard is several kilos of gold a year from the dust that falls on the clothes. After being worked, the golden parts of a watch are carefully polished, and an Asiatic gem cutter precisely reworks the costly jewels, already cut when purchased. Such a procedure results from thinking of perfection and is necessary if the hands are not to get stuck on irregularly "high" jewels mounted on a completed dial set with gems.

The bands are also put together by hand under the loupe out of many individual parts, or artistically plaited from gold wire. Here too, modern technology is not ruled out: computers are used by the handworkers to

calculate the optimal form of the smallest wheels and pinions, and by the artists to complete their exclusive designs.

Just as Chopard has become synonymous with extravagant decorative watches, the "Happy Diamonds" line has long been a Chopard trademark.

Women's and men's models in steel and gold, from the "Gstaad" line, with sweep second and date indication; cases watertight to sixty meters.

The idea for this jeweled line came from a Pforzheim designer, who created the ancestor of "Happy Diamonds" to enter in a contest, won a prize and joined the ranks of Chopard prize-winners at international competitions (fifteen "Golden Roses" in Baden-Baden, three "Diamond International Awards" in New York). There were great difficulties in making the "Happy Diamonds" line. The faceted stones all had to have their glittering cut surfaces facing forward instead of in all directions—otherwise the effect would be lost. The problem was solved and is now a company secret. But then another difficulty had to be overcome: the dealers could not imagine that "this toy" would attract customers. The basic idea, "how splendid it is to be rolling in diamonds", seemed to them inapplicable to watches. But the Scheufeles produced the watches and the customers were attracted. Today "Happy Diamonds" continues to be the firm's most important line and will be preserved carefully and expanded further.

The second Chopard line is more sporting: "St. Moritz". The small sun on the dial and the original certificate from that prominent place show that the city fathers were its godparents and gave it their blessing. The idea for the name came from Junior Manager Karl-Friedrich Scheufele. The "Monte Carlo" line has also received official certification. The principality's coat of arms thus appears on the dial. With "Gstaad" Chopard returned to Switzerland for the source of a line's name.

No matter which line the watches belong to, a detailed register is kept at Meyrin, in which the reference numbers, running factory numbers and delivery dates are documented. First of all, this helps to expose falsifications as quickly as possible and rule out a counterfeit market, with which there have been no problems as yet. For the customer this register has the advantage of making documented information available at any time as to when his watch was made and to which dealer it was shipped. Special features made to the customer's order are also noted. But even many a sheik has backed off when he heard the high price. But whoever insists on his own unique creation can even choose the design. There is only one limitation: it must harmonize with the style of the house.

Man's wristwatch with perpetual calendar and leap-year indication.

Man's wristwatch with automatic winding, day, date and moon-phase indication; drawing by David Penney.

Man's gold wristwatch with automatic winding, date indication, chronograph, 30-minute and 12-hour indicators. The dial has an additional tachometer scale. Screwed-on winding and hand-setting crown; 1988.

Man's gold wristwatch with "Eternal" calendar, moon phase, leap year and 24-hour indication, new model of 1988. The watch is regularly sold with automatic movement, and also with quartz movement on request.

Wristwatch with chronograph, model "Mille Miglia", made in steel, watertight to 100 meters, automatic movement.

Upper left: man's wristwatch with moon-phase indication, "Luna d'Oro" line.

Upper right: classic eliptical model for men.

Center: man's gold wristwatch set with diamonds, from the "St. Moritz" line.

Lower left and right: models with gold bands from the "Monte Carlo" line.

chance. On the eve of the opening of the Basel Exposition it was found that several dials for the new samples could not be finished on time. In place of dials, simple gold plates (with the Corum emblem at the 12 position) were put in. The customers were interested, and the "solution of necessity" set a new style trend for the watch industry.

In 1974, though, came the final breakthrough with the "Coin Watch" that made Corum world-famous overnight. Since then it has become practically a "must" for every American president.

This line, in which the movement is "hidden" in an original coin, began with the American "double eagle" twenty-dollar gold piece. But soon these watches were also made of Florentiners, Pesos, Napoleons, Vrenelis and Krügerrands. But all models have one thing in common—a valuable coin, often available only as a collector's item, in mint condition and perfectly worked. The coins are cut through the middle by a Swiss specialist using a secret process. The upper half, enclosed in a sapphire glass, becomes the dial, the lower half is machined out to take an ultra-thin automatic movement. The milling on the edge of the coin is redone after finishing. In 1976—as technical development strode forward—diamonds first decorated the coin; from 1979 on, quartz movements were also built in, and later the watch was given a watertight case. With this "Time in Money" watch, Corum could finally make its mark, particularly on the American market, with the advertising slogan, "A little something for your great-great-grandson". But the model found fans in Japan too. Striking things have a way of succeeding.

The road to the gold-bar watch was paved: the ticking gold bar on the wrist became reality. Corum has the gold bars produced exclusively for its own use, for the guaranteed weight of the bar (also available now in platinum), the front of which forms the dial, has to reckon on the borehole for the hands.

Gaston Ries did not live to see much of these developments by his firm. He died in

Wristwatch with a dial decorated with a real peacock feather; Seventies.

Other wristwatch models from the Seventies; at right, one from the "Love Bond" line.

84 CORUM

"Black-White" version of the famous Corum Admiral's Cup watch with "Nautical Hours" dial. The colors of the marine flags are shown in the customary hatchings of heraldry. Quartz movement, water-tight, calendar near the six, gold (18k) case, glass ring set with 63 diamonds (0.10k), available as a man's or woman's watch.

With a coin watch and the slogan "Time in Money", which is still the name of this Corum line, Corum opened its greatest advertising campaign in the USA in 1968.

In its massive gold case (watertight to 60 meters), with angled lunette and hand-milled dial, this watch offers a "réserve de marche" indication (eccentric dial by the three), linked to the automatic winding system, while the reverse can show either "Helvetia" as a symbol of Switzerland, or a family crest, in fire-resistant enamel.

Model from the "Admiral's Cup" line, its dial enameled with nautical hour indications.

right
One of the newest creations is this man's wristwatch with automatic winding, date and winding indication, and steel-gold case and band.

CORUM 87

An authentic piece of meteorite stone forms the dial of the "Meteorite". The coordinates of its landing place are engraved on the bottom of the case.

The movement of this wristwatch from the "Casino Royal" line is mounted on gimbals.

1958, but the spirit of this middle-class family business, that was able to react quickly to developments of the market and fashion trends, lives on. Jean-René Bannwart, Junior, brought additional impulses into the firm as its new partner.

The products of Corum have always stood out. Whether the feather watch with a genuine peacock, ibis, pheasant, grouse or other rare feather on the dial (which can be ordered only by collectors), or the very newest creation, the meteorite watch, whose dial is made of a golden plate covered with a skin-thin slice of meteorite (half a millimeter thick). The two watches have something in common: every example is as individual as a fingerprint. For meteorite stone is rarer on earth than gold and therefore harder to obtain. An American professor arranges trades among the world's museums. His commission: a piece of a meteorite (usually from Arizona or Mexico, more expensive than gold), which is sent to Switzerland and prepared there. So it becomes possible for the customer to wear a piece of space on his wrist.

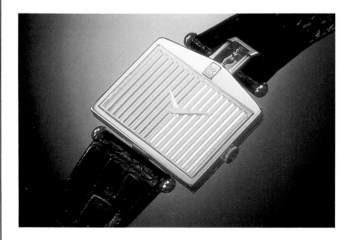

In the House of Corum the creations have to be unique. This can be seen not only in the "Golden Bridge", a "ladder" watch which, to be sure, consists of only 37 high-precision, finely chased parts, but for the wheel train and escapement of which 130 different work processs are required. Some 2000 pieces of the "Golden Bridge" have been sold to date.

The extravagant can also be found in:
— the "Admiral's Cup", with flag symbols instead of numerals on the dial (for the women's version of which a new design has just been developed),
— the ultramodern, classic "Romulus" (the Roman numerals are engraved by hand. Proof thereof: once a model with two X numerals but no XI was sold by mistake),
— the "Remus",
— the new "Westminster" (the dial is an exact copy of that of "Big Ben"),
— the "Pyramid" (with its pyramid-shaped cut glass of white sapphire),
— the "Rolls-Royce" (a unique reproduction of the radiator grille of the "Camargue" model in white gold, more expensive than the car itself), and
— the "Clipper Club", in which twelve massive gold screws mark the hours,

Thus every watch has either its own symbolism, or René Bannwart can relate its own history. All models have one thing in common: they are usually made in three sizes, as women's, so-called medium, and men's watches, for the very large men's watches are popular almost exclusively in the German-speaking countries.

About ten thousand watches (most with quartz move-

The radiator of a Rolls-Royce inspired the design of this man's wristwatch.

The wristwatches of the "Golden Bridge" line, with their "ladder" movements, belong among the "classic" Corum models.

ments) leave the factory every year, made in all types of gold (red gold, white gold, yellow gold), in steel-gold, blue steel-gold, steel and platinum. About a third of them are women's watches. Corum values the fact that the raw movements by ETA (quality level one, from LeCoultre and Piguet) are finished in their own workshops by masters of their trade, who supervise the construction of all models, and treated with a special oil. The only exception is the "Golden Bridge", which originates completely in its own computer-guarded house, always in a series of a hundred pieces, with an assembly time of three months.

Corum's worldwide bestseller is the watch worn by the King of Jordan: the "Admiral's Cup". The coin-watch, of course, is at the top of the American "hit parade", followed by the "Admiral's Cup" and the "Romulus". The last-named classic wristwatch is also the most popular in Germany. The new star could be the "Meteorite". Two hundred were to be made in its first year of existence—five hundred were made. And another five hundred orders are already on hand for its second year.

Extravagance as the firm's characteristic pays off. The only complaints at Corum are about the "penny-pinching measures" of the owner who has a battery changed by a low-price dealer instead of a specialist. But even when a "cheap Charlie" or the owner himself causes damage through carelessness, the factory can still help. Not only because every watch is registered, but because the origin of all the individual parts can be found on countless index cards (the computer will take over this job only gradually). And if some part is missing in the spare-parts warehouse during repairs, then it will simply be made. René Bannwart puts it like this: "We always treat every watch as carefully as we do when we make it."

And that is a promise that is already applied when a new model is developed. The final design is simply approved by the chief when he writes "O. K. Bannwart" on a white sheet of stationery with gold letterhead. Then the same artists create the prototype. And that too is something that Corum makes unique, but that well suits Bannwart's business philosophy: "We are a family. Only when everyone enjoys his work do we get a perfect product. Every one of my colleagues can come to me at any time."

And the colleagues make use of that. That is why Corum quit work for the day at 3:00 P.M. when the newly-elected Swiss president came to La Chaux-de-Fonds for the first time. The family wanted to see the politician, and the family council agreed.

Ultra-thin models from the "Romulus" (left) and "Remus" (right) lines.
A fifteen-gram gold bar from the Swiss Bank Company forms the case of this man's wristwatch from the "Les Spéciales" group.

EBEL

Two decades ago nobody knew the German tennis star Boris Becker. It is no wonder, for the super-sportsman had just been born. But two decades ago scarcely anyone knew the "Ebel" watch brand, the brand for which the youngest Wimbledon winner does commercials.

The amazing rise of an almost unknown watch manufacturer is closely linked with a single person: Pierre-Alain Blum, a disobedient severed and made progress by using his stubbornness. But before we report on the manager, born in 1946, who catapulted the company, which does not yet have a century of history behind it, into the leading group of Swiss watchmaking firms, we must first say something about his grandfather.

Eugène Ebel-Blum founded the Ebel factory in La Chaux-de-Fonds in 1911, forming the name from "Eugène Blum et Lévy" (the last is the grandmother's maiden name). The factory worked then as an assembler: watch components were purchased and built into complete watches. Sometimes the name "Ebel" was on the dial, but usually the name of a contracting firm. The founder's son did not change this business system much. He led the firm through the difficult Thirties and Forties, expanded the distribution network, and stressed high quality. It was only the grandson, Pierre-Alain Blum, who made of Ebel a concept that included a portion of good luck, as he himself admitted. But was it really only luck?

He was—and makes no bones about it today—a poor student. One Sunday night when he was fifteen, he marched into his parents' bedroom and simply announced that he was not going to high school any more. His parents were probably shocked, but from then on Pierre-Alain attended a technical school. He wanted to be a mechanic or engineer. The watches of his parents and grandparents meant nothing to him. For four years he studied electricity, and still had two years of study ahead of him when he announced: "I'm not going to be an engineer either." He decided to take a basic course in watch technology, came to La Chaux-de-Fonds, but quit after two months. Sitting behind a workbench for the rest of his life was not for him. Instead of that, Pierre-Alain Blum wanted to prove to his parents that he could stand on his own feet. In 1964 he packed his bags and traveled to New York, found a job with Lucien Picard, a small firm that sells Swiss products in New York, and specialized in the sale of watches. From then on he worked his way up. When a colleague left, Pierre-Alain took over his work. After five years of work, the yearly income of the firm had risen from three to twenty-one million dollars. At Christmas of 1969 the "boss" offered the twenty-four-year-old a partnership. The young Swiss wrote this to his father. "Come home", came the answer, "and help me manage the family business, otherwise I'll sell out."

Eugène Ebel-Blum. Man's wristwatch with automatic winding from the Fifties.

"Go ahead and sell!" Pierre-Alain Blum wrote back, but changed his mind when a single sentence was telegraphed back from Switzerland: "Blood is thicker than water."

The "prodigal son" returned to Switzerland. In the States he had already earned a hundred thousand dollars a year. His father's business amounted to ten thousand. He explains his decision thus: "If somebody else had bought my father's business and made a success of it, I would have been an idiot. If he had failed, I would have said he was an idiot and been annoyed with myself that I hadn't tried it."

At first he was not happy with his decision. He was the "son of the boss", could work day and night, bring in a gigantic contract from the States—and his father alone made the decision to take on only 10% of it and decline 90%. Or there was the time when Pierre-Alain traveled to Venezuela and came back without an order. His father was furious.

On account of a serious accident, his father had to withdraw from active participation. That was Pierre-Alain Blum's chance. He could now reign and rule as he pleased. The results: a 30% increase in income the first year, 30% the following year and twice as much the year after that. In 1973 the son bought 70% of the business from his father; two years later he owned the entire firm. The rise continued. From 1970 to today, the work force increased from forty employees to about seven hundred. The income multiplied by the factor 61 in fifteen years. In the firm's seventy-fifth anniversary year, an income of about 65 million Swiss francs was to be seen in the ledgers.

But it was not just Pierre-Alain Blum's business sense that, for example, made him decide to travel to Hong Kong and Singapore to establish branches there. The success of Ebel is based not only on capable marketing and striving to sell watches of the highest quality, but also the name of Cartier. Early in the Sixties the Paris jeweler looked for a manufacturer to make his luxury watches. Cartier and Ebel collaborated so well that at the present time nine out of ten dials that are set on movements at Ebel bear the Cartier signature. That was the reason why the firm's income suddenly leaped upward. At the same time it also caused worries for the firm's owner, for Ebel became dependent—some-

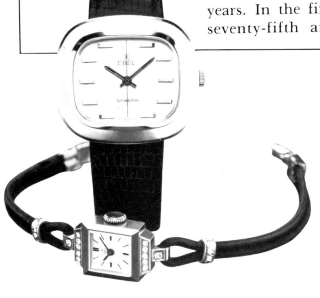

A woman's diamond-set wristwatch from the Forties and a man's model from the Sixties.

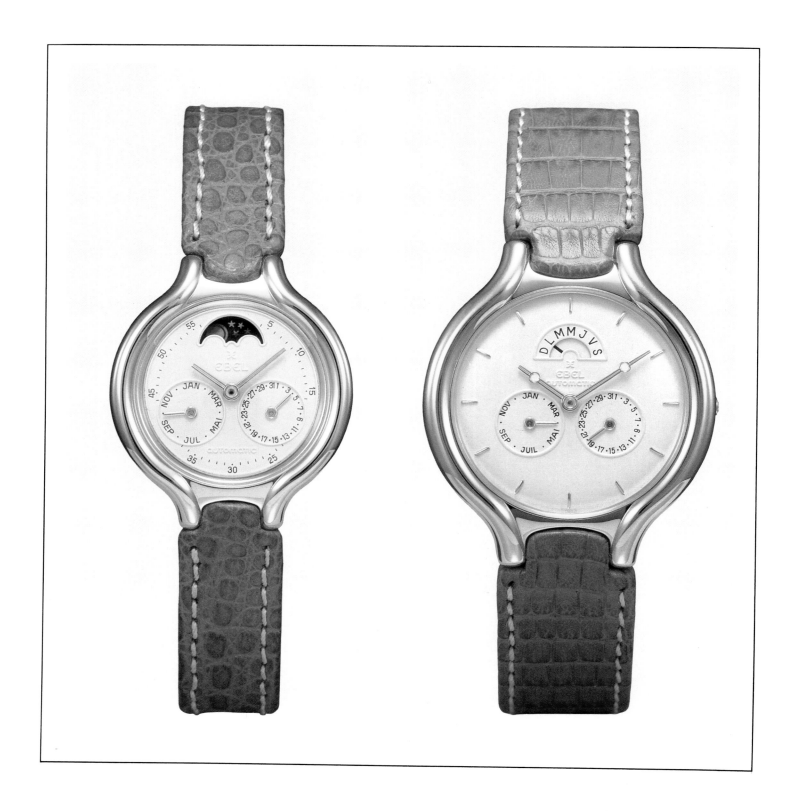

Two models from the Beluga line with automatic movements and a 4-year calendar that needs correction only on February 29 of leap year; gold cases.

At left a woman's model with date, month and moon phase indication.

At right the man's "Agenda" model with day indication in place of the moon phase.

Man's wristwatch with perpetual calendar, leap-year and moon-phase indication, from the "Beluga" line.

94 EBEL

Man's classic gold wristwatch with sweep second and date indication, from the "1911 Collection".

Man's wristwatch from the "Discovery" line in steel/gold construction; the case is watertight to 200 meters. The dial is legible at night; quartz movement with date indication. The lunette can be turned only counterclockwise for safety reasons.

thing inconceivable for the dynamic businessman.

So as not to have to give up his own production completely, he looked for a new self-concept. The Swiss watch designer Edy Schoepfer was given a contract to design a "typical Ebel". He created the "Leader", that elegant sport watch that originated the elegant "Ebel look", recognizable by the flowing transition from metal band to case, and by the five gold screws on the attachment. Priced in the lower range of luxury wristwatches, it became a "coming man's" watch, whose design has not changed to this day except in becoming somewhat thinner through technical improvements.

The "only" thing the firm needed was customers. Blum "invented" sport sponsorship in the watch industry when his friend Alain Prost wore an Ebel watch in the French Grand Prix, thus calling attention to the brand. Advertising expenses eat up at least fourteen million Swiss francs a year, as experts estimate. In any case, he created thereby a profitable degree of exposure for the Ebel brand. Sport sponsorship was extended to auto racing, golf tournaments, sailing regattas and tennis tournaments. Boris Becker is under contract to Ebel, and so are Yannick Noah, Sandy Lyle, Markus Ryffel and Valérie Brisco-Hooks. This has meanwhile been joined by the "furthering of culture" (concerts with star conductor Leonard Bernstein, opera performances and ballet evenings with top artists), which has created esteem for and confidence in the firm and its strategy. This confidence is also seen in Ebel's model and price policies.

The firm tries hard to have the same models sold for same prices worldwide, fights off discounters

Women's and men's wristwatches in steel-gold, from the "Leader" line.
Model with date indication, from the "Beluga" line.

in the USA, and is modest about new models. The "Beluga", now almost legendary, which was Ebel's second great success, was its only new model for five years.

"We can sell more watches that we build—so why should we make a new watch?" Watch owners are happy to see this, for when a line remains unchanged, the value of the individual watch will never fall, but will rise steadily.

So along with the "Leader" (steel and yellow gold with two-tone band, date and second indication, watertight to three atmospheres) and the "Beluga" (gold, ivory-colored dial, watertight to two atmospheres), there are now the "Discovery" (quartz movement with date and second, indication that a battery change is due, turning bezel, watertight to a depth of 200 meters, steel and gold), the collection piece (automatic chronograph, yellow or white gold, watertight to three atmospheres), and the chronograph with perpetual calendar and moon-phase indication (yellow gold, watertight to thirty meters).

The "1911" is new, a replica of an old Ebel, meant to illustrate the continued philosophy of the "architect of time": thin is out, the "fat", solid watch is back in demand.

All Ebel watches are constructed in the firm's own workshops. About 300,000 pieces leave the factory yearly, 20% of them with automatic movements (Beluga, chronograph and perpetual calendar). Where today cases for Ebel and Cartier are combined with raw movements by other manufacturers (including Piguet) at neighboring tables, the Ebel raw movement will soon originate. For Ebel is working on the development of its own quartz caliber.

Technology is utilized particularly in quality control, contract work (computers finalize the production plans and deal with delivering individual parts) and designing, while handwork is wanted in refinement. For example, the five gold screws of a "Beluga" are installed by the hands of a master. That must be so, they say at Ebel; a machine would only scratch the sensitive lunette, and such a valuable watch just can't be made without fingertip feeling. This is also true of the crown of that model, and the reinforcing rings of the diver's watch. When the computer-tested running regularity is known, a master must take the watch in hand again, for it will be regulated according to handworking tradition. The testing of running regularity takes the factory about three weeks. Then reference and movement numbers are noted, as well as the agent who receives the watch. The latter must keep a record of its retailer, who in turn must record the purchaser's name. In this way Ebel wants to shut out the counterfeit market in advance and keep an eye on business. The procedure for antique Ebel watches is different, but the year of construction can still be learned from the factory.

Ebel is well aware of the fall of crude oil prices. The Near East, still the market with the highest number of sales from five to seven years ago, has fallen to third place behind Europe and the USA. But Pierre-Alain Blum, who does not know whether the two sons of his first marriage will want to carry on the family business, has no sleepless nights about it. It is his goal to assure success and keep growth under control. New models would only cause disturbances.

Along with a chronograph, this man's gold watch with automatic winding also offers a perpetual calendar, leap-year and moon-phase indication.

Wristwatch with automatic winding, chronograph with 30-minute and 12-hour registers and date indication, gold case.

GERALD GENTA

This chapter really should begin with the words, "Once upon a time there was...", for the artist who, without a doubt, produces the most remarkable—one might even say the wildest—wristwatches in the world today, has had a fairytale career, has made a dream come true, has taken an idea with which he has been possessed all his life and turned it into reality.

At the age of fifteen, Gérald Genta, whose parents emigrated to Switzerland from Italy, began his apprenticeship with a jeweler in his birthplace of Geneva. He ended it four years later, created advertisements as a graphic artist, took a trip into "haute couture", where he learned the art of proper cutting, and suddenly recognized his love for watches. He followed his inner voice, created watch designs, and soon became the "number one" designer in the world of Swiss watches.

Watches that everyone knows today, and that are still successful on the market, go back to his designs:
—the "Royal Oak" of Audemars Piguet,
—the "Nautilus" of Patek Philippe,
—the "Titan" of Omega, and
—the "Bulgari" of Bulgari.

For twenty years Genta pursued his goal of creating ever-new forms and clear lines. There is practically no Swiss luxury watch manufacturer who has not had Genta design one or even two case forms. But his masterpieces were never sold under his name. This "silent" creation finally seemed to come to a sudden end when in 1968 the Geneva manufacturer, Universal, was honored with the "Diamond International Award" in New York. It was also a great success for the designer, Gérald Genta, but one that brought him more frustration than joy. For once again his name was not mentioned. After this bitter, disappointing experience, he decided to step out of anonymity: from then on his watches were to be known by and sold under his name. But this took place only four years later, because the artist first had to fulfill contracts with Universal and Audemars Piguet.

The time finally came in 1974. Genta bought two small, unknown watch factories in Geneva and Le Brassus. He first had the buildings painted antique pink, so as to stress the "Genta style" from the very start. He hired watchmakers, engravers, stone-setters and jewelers (today there are eighty employees, including forty watchmakers and twenty jewelers), and the adventure of independence could begin.

But who or what was Gérald Genta then? A name that the insiders in the watch industry knew, but one that still had to find a place as a watch brand. The beginning was not easy for the "Picasso of watches", even when the first orders from Fred in Paris and Van Cleef & Arpels in New York arrived in Geneva. The particular feature of the "Fred" watch was a hand-engraved dial whose parallel lines created special light effects.

In those years Genta production followed two tracks. For one, watches for other firms,

Gérald Genta

One of the most complex and most expensive wristwatches of its time: a model with minute repeat, perpetual calendar, leap-year and moon-phase indication.

whose names appeared on the dials, while his signature was engraved only on the movement; the other, a number of individual pieces which he created to suit the wishes of a sultan, a king or a head of state who wanted to wear a very unique watch. These individual creations, which Genta made to the wishes of his customers or—when he had a free hand—made so that the wearer's personal characteristics were given artistic expression in the appearance of a wristwatch, are still a specialty of the house:

—a recent work is the "Dracula Watch", in which the rubies on the dial represent drops of blood;
—a "Scorpion Watch", which he built so that a constellation of gold is set in pearls and the claws indicate the time;
—a watch with a large precious stone set over the dial instead of a crystal, causing a firework effect of colors.

There are many small details with which Genta can meet the special requests of his customers at any time. A wristwatch with minute repeat was equipped with red numerals and hands (the hands, naturally, were made of red-painted gold), because the client had poor vision and could recognize the color red most easily. One of Genta's special services is that the master wears the finished watch himself for two weeks to test it before it is delivered to the customer.

Then as now, handwork has been a support to success. Engraving the signature and reference number on the movement is done so that they are hidden. Royal arms and Mickey Mouse figures are cut out of gold, engraved and painted by hand.

Just the cutting of a small coat of arms for a dial can take more than two hours.

Soon after the founding of his own factory, a Japanese customer paying a visit to Genta said that he represented the "Spirit of Geneva" to those who knew watches. This sentence stayed in the unique watch producer's mind. In 1978, shortly before he finally moved into Geneva, he made these words his motto: "Gérald Genta, l'Esprit de Genève".

The "time-artist" was not satisfied with only remarkable forms (glasses-wearer, goatee). He also wanted to offer the best technically, and demanded the finest achievements from his watchmakers and suppliers, making it possible to give lifetime guarantees with Genta watches. He proved that what he promised was not just empty words with a calendar watch that he gave to an expedition to take along to the 6959-meter high Mount Aconcagua in Argentina. Eighty-five years after a Swiss team had first conquered the highest mountain in the Americas, this watch came through all the difficulties.

It withstood all the many back-

"Secret Time", a man's one-handed gold wristwatch.

Man's wristwatch from the "Safari" line, with day, date, moon-phase and 24-hour indication. In the clasp of the band is a compass.

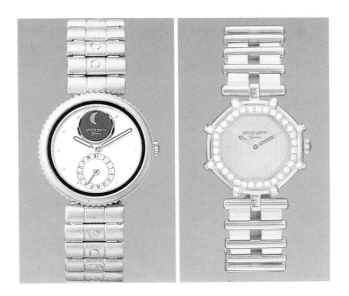

and-forth movements while driving hooks into the rock, it maintained its precision through extreme temperature changes—the watch was unharmed, the sapphire glass unscratched. The battery of the quartz movement was located so that it was warmed by the wrist. A gold decorative watch that cost a small fortune withstood pressures that many a steel sport watch never faced. The built-in moon-phase indication enabled the mountain-climbers to be ready for the weather changes that they knew would take place with the full and new moons. After the expedition, Gérald Genta simply had the ornamental watch polished anew.

The relationship of mechanical watches to quartz watches has evened out at 50:50 today, as has that of women's to men's watches. About 2600 watches a year are sent out to all over the world from the Rue de St. Jean 19; Italy, Asia, the Near East and England are the most important markets. Genta tried a move into the USA in 1986, after his watches had found more and more admirers there by roundabout ways.

The division into individual production of single pieces and series production of regular lines has remained. But sometimes single pieces are developed into new series: for example, from a wristwatch with a horse's head surrounded by many gemstones there grew the "Menagerie" line, with swans, snakes, lions and other animals.

Another example is the "Gefica Safari", that was "born" in Africa when three hunters complained about the insufficiencies of their watches as they sat around their campfire one night. One of them later turned to Gérald Genta for advice. His answer was the water-tight "Safari" with alarm, second time zone, moon phase, day and date, with gold or sharkskin band and a compass on the clasp.

The "L'Open" also is based on clients' special requests; on it four golfers can keep a

The "L'Open" with its four scoring registers is intended for golfers.
"Gold & Date", a man's octagonal gold wristwatch with date indication.

Gold-band watch with date and moon-phase indication, from the "Safari" line.
"Gold" line, bezel decorated with diamonds.

102 GERALD GENTA

These ornamental wristwatches are richly set with diamonds; their dials are protected by cut semiprecious stones.

record of their strokes via four different push-buttons. The scores are shown digitally on small registers, and can be set back to zero by using the central button. Genta does not make everything alone. Movements or complexities are made in Le Brassus. The raw movement for a technical masterpiece, the unique automatic watch with minute repeat, comes from LeCoultre or Piguet. The hands come from another supplier. But the cases, bands and, as happens very rarely in the business, the dials are produced and prepared in the firm's own house. Only a part of the stone-setting work is contracted out to Geneva jewelers. The quartz movements are "trimmed" for highest performance with Genta's own mechanical complexities. A register with reference numbers and notations of the supplied retailers (150 to 200 worldwide) is traditionally kept on index cards for every watch; information is provided in writing.

The really classic lines today are the "Tropics", the "Gold & Gold", the aforementioned "Safari" that is also made in bronze, and the likewise noted "L'Open". The most expensive series pieces are the automatic watches with perpetual calendars and/or minute repeat (prices go to about 200,000 Swiss francs).

In a tan-painted safe in the video-surveilled house, which can be entered through the cellar door and after passing a heavy blue metal grid, lurk expensive individual pieces, such as a wristwatch whose crystal is set with 11 karats of diamonds. The price: a million Swiss francs. The corresponding women's watches with diamonds and sapphires were sold shortly after the appearance of an advertisement in an international magazine.

In almost all classic Genta watches, from "Les Fantasies" on (with hand-painted Mickey Mouse figures or the Pink Panther, whose paws are the watch's hands), the client himself can decide the types of complexities, including the perpetual calendar with leap-year and moon-phase indications.

Switzerland's most creative watch manufacturer spends much time in Monaco, where he lives with his family. His goal is to attain a new high point every year. This should not be difficult for him, since he makes some 250 designs a year, of which only a few hundred have been realized. Gérald Genta will always hold a special place in the history of Swiss watches. How could it be otherwise, with so much courage to create the most remarkable designs.

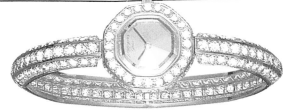

Gold wristwatches from the "Les Fantasies" line (Mickey Mouse, Pink Panther).

This ornamental wristwatch belongs to the "Menagerie" line.

"Champagne" ornamental watches for women, with cases and bands set with diamonds.

GIRARD — PERREGAUX

It was recorded that in 1791 Wolfgang Amadeus Mozart was buried in Austria, Friedrich von Schiller was Professor of History in Jena, and the nineteen-year-old watchmaker Jean-François Bautte settled down in Geneva and opened his own factory, the third oldest in Switzerland.

The young businessman, who put the main emphasis on the unique and striking appearance of his products, found well-to-do clients above all in the princely courts. His clientele grew, and it was noted that only masters of their trade really went about their work seriously, no matter whether they worked as watchmakers, engravers, chasers, enamelers, goldsmiths or stone-setters.

Since he had no blood heirs, Jean-François Bautte joined forces with the watchmaker Constantin Girard-Perregaux, who opened his own factory in La Chaux-de-Fonds in 1856, after the founder's death, giving it his own name. In the past century this firm not only built a legendary pocket watch ("Tourbillon with three bridges"), but also helped the wristwatch make its breakthrough—in the German Navy. Bismarck had just made public a secret defense treaty between the German and Austrian Empires (the "Zweibund") in 1880, and the ministry in Berlin invited a long list of Swiss watch manufacturers to introduce watches that ship's officers could wear on their wrists. Constantin Girard-Perregaux won the prize and began to produce gold watches in La Chaux-de-Fonds that could be worn on a chain band and had a gridlike metal cover protecting the dial (a design that was "reinvented" by another designer a few years ago). A special feature really noteworthy at that time: the movement of the naval watch had a small second hand.

Thus the production of wristwatches in small series began. The watch, whether a pocket or wrist type, was really a luxury article at that time because of its being worked in gold. Even today it is gratifying to see that "GP" watches were worn by many crowned heads: Napoleon III, Queen Victoria, King Farouk and many others.

In 1957 Girard-Perregaux made the jump to the automatic wristwatch with its own system. The heart of the watch consisted of two so-called gyrotrones, toothed wheels each mounted in seven ruby bearings, that used every movement of the swinging weight in one direction to wind the watch via an original mechanism.

In 1961 a change in the philosophy of the product was seen. The first electric wristwatch with a visible battery was a milestone in the development of watch movements, though its significance was not yet recognized at that point. In 1965 the first high-frequency wristwatch, with its 36,000 beats a second, created a sensation. It was a system that attained ultra-precise running regularity. The HF Chronometer (HF meaning High Frequency) came on the market as a hand-wound or gyromatic model with or without

Francis P. A. Besson

The high-frequency Model "HF" wrist chronometer, with automatic winding and date indication; latter half of the Sixties.

GIRARD-PERREGAUX 105

Foreground: the "Equation perpétuelle" model with quartz movement and mechanical perpetual calendar (showing date, month and leap-year) as well as moon-phase indication.

106 GIRARD-PERREGAUX

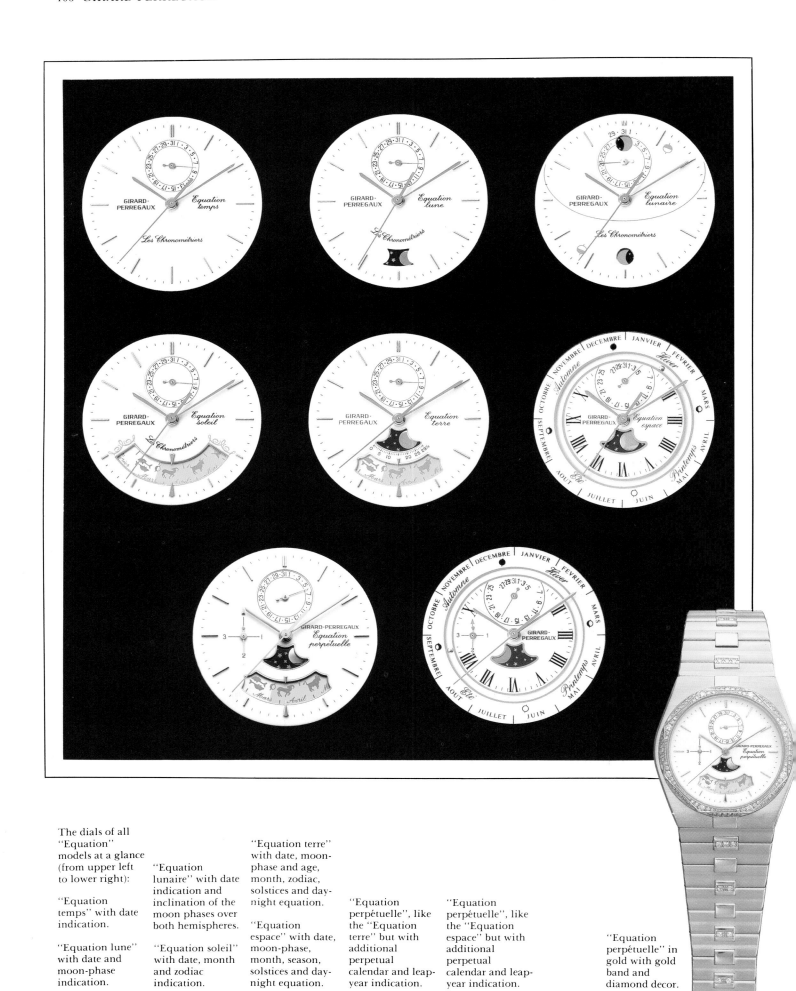

The dials of all "Equation" models at a glance (from upper left to lower right):

"Equation temps" with date indication.

"Equation lune" with date and moon-phase indication.

"Equation lunaire" with date indication and inclination of the moon phases over both hemispheres.

"Equation soleil" with date, month and zodiac indication.

"Equation terre" with date, moon-phase and age, month, zodiac, solstices and day-night equation.

"Equation espace" with date, moon-phase, month, season, solstices and day-night equation.

"Equation perpétuelle", like the "Equation terre" but with additional perpetual calendar and leap-year indication.

"Equation perpétuelle", like the "Equation espace" but with additional perpetual calendar and leap-year indication.

"Equation perpétuelle" in gold with gold band and diamond decor.

date indication and second stopping. Considerations of changing the firm's policy led back to the original roots of independent manufacturing. The firm wanted to achieve the highest possible technical standards. For that purpose, a research laboratory for the development of quartz watches was established successfully in 1966.

Switzerland's first industrially produced quartz watch was built in La Chaux-de-Fonds in 1970. The oscillator frequency of 32,768 Hz (Hz = swings per second) was also accepted as the standard by other manufacturers. The developmental work was continued, and Girard-Perregaux thus became the quartz expert of the Swiss watch industry:

— 1973: the factory produced the GP 641 movement, at that time the smallest-volume quartz-analog movement in the world (3.7 mm high).
— 1977: 80% of all tested Swiss quartz chronometers were made by Girard-Perregaux.
— 1978: another quartz mini-movement appeared, smaller in volume and diameter than a watch battery still in use in 1970.

The new orientation of the factory is attributable to its general manager, Francis P. A. Besson, a brilliant technician who settled on quartz as the swinging element of watches as early as 1961. Nevertheless it is not the opinion at Girard-Perregaux that the firm should be regarded simply as a producer of quartz watches. This self-concept is based on the facts that the spring was merely replaced by the battery, the balance by the quartz and the escapement by the electric motor. Everything else is mechanical in the best watchmaking

Mounting of the moon disc in an "Equation" model.

The famous pocket watch with tourbillon, with its wheel train mounted under three golden bridges.

"Equation intégrale" model with date indication.

Gold wrist chronometer with automatic winding, chronograph with 30-minute and 12-hour registers as well as date and moon-phase indication.

tradition. Or otherwise stated: electronics has been integrated where it is useful. Who, for example, would want do do without computerized brakes in an expensive luxury car for tradition's sake?

General manager Besson says of Girard-Perregaux: "We have brought micromechanics and electronics into a harmonious union and thereby reinvented the art of watchmaking."

One can build watches for life, several thousand pieces per year, a third of them as women's watches, and at the same time produce raw movements for other watch manufacturers in the upper price bracket. But all the same, fifty purely mechanical watches are produced per year. The markets for GP products are Europe, the Middle East, the Far East, and North America.

The exclusivity of a Girard-Perregaux is underlined by the fact that only the smallest series are made. A series of 500 to 1000 pieces requires about ten months of preparation. The twenty-five variations of the four house calibers make 20,000 finishing plans necessary, including everything involved. The research department is also exclusive, a luxury in Swiss manufacturing, but one that is afforded in order to remain independent. This independence bears fruit in the firm's self-developed modules with microwires less than two hundredths of a millimeter in diameter, or in their own micromotors (14,000 wrappings of extra-thin copper wire). Tools and testing instruments are largely made in the firm's own workshops, with a few parts made by suppliers to the firm's specifications. The only standard element is the battery.

All bridges, mechanisms and screws are hand-polished and ground on the edges; every movement is individually covered with rhodium or gold; even the oiling channels around the ruby bearings are polished. Absolute quality requirements apply in the final control: nitrogen bath (-170 degrees Celsius), heat box (95 degrees Celsius), acceleration measurement with approximately 1000 G—The movement alone is tested for nine days, the complete watch for four more days. Standing over it all is the "Department of Maintaining Quality and Reliability", which can "mix into" any department of the firm at any time, and the General manager Francis P. A. Besson, who is responsible for all the factory's products.

The complete reconstruction of a "Tourbillon" from the last century was for him only a small example of Girard-Perregaux' ability to do anything, even building a watch in a small series (twenty pieces), with only one piece as a

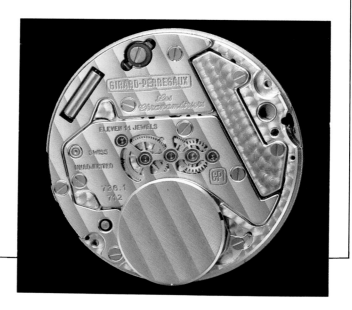

Man's wristwatch with automatic winding, chronogram with 30-minute and 12-hour counters plus date indication. The tachometer scale engraved in the lunette is noteworthy, 1988 model.

Movement and under-the-dial view of an "Equation" model.

model and no surviving plans or assembly system. Fourteen of these watches (priced at a quarter-million marks apiece) were sold, two to Germany, the rest in Switzerland, England and the Middle East.

The firm's present-day philosophy can be seen in the "Equation" line:
— Synthesis of microelectronics and micromechanics.
— Synergy (the microelectronics gain from the refinement of the watch via traditional handwork).
— Simplicity, because all models, despite some of their functions being highly complex, are controlled only by the crown.
— Symbiosis of time and man. Analog indications, ergonomic form design, transparent case bottom.

Principle: one sees not only the time that passes, but also that which is already past, still remains, and is yet to come...

The various models are:
"Equation temps" with second, minute, hour and day indication as the basis of all Equation models;
— "Equation lune" with additional moon-phase indication;
— "Equation lunaire" with moon-phase indications for northern and southern skies;
— "Equation soleil": instead of moon phases, a view of the zodiac with the colors of the seasons;
— "Equation terre": indications of moon revolution time, moon phase, zodiac, month, season, winter and summer solstices, day-night equation, in addition to the regular indications;
— "Equation espace": a small gold globe of the earth circles the dial in a year, showing months, seasons, winter and summer solstices, day-night equation;
— "Equation perpétuelle" in two versions (basic format that of either "Equation terre" or "Equation espace"). In addition the perpetual calendar with leap-year indication, which like all the functions of an "Equation" watch, is always switched mechanically.

The precision and love of detail that Girard-Perregaux devotes to its watches can be made clear by two examples:
— The disc with the small diamond-cut gold earth-globe in the "Equation espace" is only two-tenths of a millimeter thick; and
— every dial of sapphire is decorated with gold.

Francis P. A. Basson says of that: "We have done a lot of work so that our customers will have nothing to do but treasure the results of our efforts."

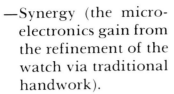

Schematic drawing of an "Equation perpétuelle".

The "Equation perpétuelle" shown schematically at the left, as a wristwatch with gold case and leather band, as well as perpetual calendar and leap-year indication.

IWC

You can twist and turn it any way you want: it began with a bankruptcy! And yet: If this American, Florence A. Jones, had not existed, those who love watches would have to do without the products of the International Watch Company—IWC for short.

In the past century this F. A. Jones had worked in the growing American watch industry, at the factory of George P. Reed in Boston, Massachusetts. One day he got the idea of having the pocket watches that were so popular in America produced in Switzerland. This decision was based on solid economic considerations, for salaries in "good old Europe" were considerably lower than in the New World. So F. A. Jones traveled to Switzerland and got acquainted with the Swiss businessman Johann Heinrich Moser in Schaffhausen.

This J. H. Moser was a watchmaker, had produced and sold watches in Petersburg, the present-day Leningrad, founded a factory in Le Locle, and constantly tried to use the water power of the Rhine to build up industrialization in his home town, Schaffhausen. But the city fathers were not eager to consider his plans. There were still small firms in Schaffhausen that drove their machines by big water wheels. But only when a drought sharply lowered the water level and put the factories out of action did responsible parties follow Moser's ideas and have a large dam built. The Rhine water now flowed through big turbines that transmitted their motions to the factories via gigantic pulleys and lines. Now energy was available in large amounts, but there were not enough factories willing to move into Moser's "industrial zone". At this moment Jones appeared on the scene with his plans. He and two other Americans had just founded a company in New York for the not yet manufactured watches: the "International Watch Company", with the purpose of producing watch components and finished watches in Switzerland exclusively, to be imported via New York and sold in the United States.

New factories could now be built and connected to the "Moser Water-power Works", as production was planned from the start in large quantities of at least ten thousand pieces a year. In America an advertising campaign began with the picture of an imposing factory, which existed to that extent only in the imagination of Florence A.

Johann Heinrich Moser (1805-1874).

View of the IWC factory building from the north.

Jones. The same could be said of the predicted quantities of watches, for less than half of the planned watches were produced after the founding of the firm in 1868 and signed with the signet, "International Watch Company", and delivered to New York. The reasons were that building the factories had taken longer than had been expected and operating the machines made more problems for Mr. Jones than he had anticipated. The money ran out, and Mr. Jones looked for new partners.

The first attempt promptly went awry. Suddenly the firm belonged wholly to F. A. Jones. With the help of the Handelsbank he then tried to form a stock company. This company not only took over the firm with its 170 workmen, but began to build additions. But mistrust was aroused among the stockholders, and as 1875 began, an investigating commission was formed. Thus it was learned that, instead of 10,000 watches, only 4072 had left the factory within a year, and that Jones had given a 220,000-franc contract for a new factory building. The results were a bankruptcy action and the flight of Jones. That was the end, or should have been.

But the Schaffhausen Handelsbank prevented the collapse, took over all the assets and named an American, Ferdinand F. Seeland, as manager. He was not unknown in the American watch industry, but he spoke neither German nor French. Still he succeeded in getting the business moving again. He introduced new types of movements, conquered not only the American market but the English, Russian, Austrian and German as well, showed a profit for the 1877-78 year, but disappointed the stockholders in his third year of running the business.

Quote: "We have the honor of presenting to you, as befits your position, the reckoning and the business report for the year 1878-79. But before we discuss this presentation, we unfortunately find it necessary to make you aware of an event that has become fateful in the highest degree for our corporation. As you are aware, our manager F. Seeland secretly betook himself away from here at the beginning of last August, just at the time at which the inventory had to be taken, and in the absence of the board of directors and without any word to the authorities of the company, took a trip to America with his family. After our director, Mr. Johann Rauschenbach, had received the first news and hurried home from Gastein, where he was taking the cure, the directors immediately circulated to all the friends of the business a circular to the effect that Seeland had been relieved of his position at the factory."

Johannes Rauschenbach, himself a stock-

Man's rectangular model, gold case with gold link band, barrel-shaped movement, sold in 1934.

Man's silver wristwatch with enamel dial, sold in 1914.

Gold wristwatch with hinged case, sold in 1916.

holder, audited the books and again declared bankruptcy. But he awakened the International Watch Company to a new life by buying the firm, which was valued at 592,500 francs, at an auction for 280,000 francs. It was able to go on...

For some years the company remained under family ownership. In 1895 it was reorganized so that all watch components, hands included, could be produced by the firm, and in 1897 the change was made from water power to electrical energy.

Under Seeland, not only had the "Jones Caliber" been developed, but so had new, simpler models, which were intended, along with new case designs, for low-price sales on the American market. This became a necessity when the U.S. Government placed a 25% protective tariff on watches not made in America, in order to safeguard domestic industrial production. This wiped out the advantages to IWC of the low salaries in Schaffhausen.

The firm introduced a noteworthy series of models in 1885: pocket watches with digital time indication. Tsar Ferdinand I of Bulgaria obtained one of them to give to a politician in his country. And there were even models with Chinese lettering, which shows how large the market for IWC watches had grown by then.

In 1893 a new phase of technical progress began for the firm with the Caliber 52. The movement was formed so that it could not be understood at a glance by traditional watchmakers, and the firm had to add detailed descriptions to their catalogs until 1911. Even before the turn of the century, the 12.5-ligne Calibers 63 and 64 had been developed, with which the firm was able to meet the demand for wristwatches that had increased strongly at the beginning of the Twentieth Century.

The firm quickly dealt with World War I by making a wristwatch with enamel dial and luminous numerals for soldiers. It had a lid that was decorated with the portrait of Field Marshal von Hindenburg. After the end of World War I the firm, "Watch Factory of J. Rauschenbach's Heirs, formerly International Watch Company", which had meanwhile specialized in the production of women's watches, again hit bottom, when under Ernst Homberger-Rauschenbach (the new proprietor) the sales market was limited to Germany and Austria. Exports to the USA, which were beginning again, took a bad fall with the stock-market crash of 1929 and were only revived through cooperation with Patek Philippe. A modernization of the machine shops resulted, which helped IWC to handle the economic difficulties of World War II.

World War II also brought new

Hunting cased wristwatch with enamel dial, picture of Field Marshal von Hindenburg on the spring lid.

Man's gold wristwatch, 1938.

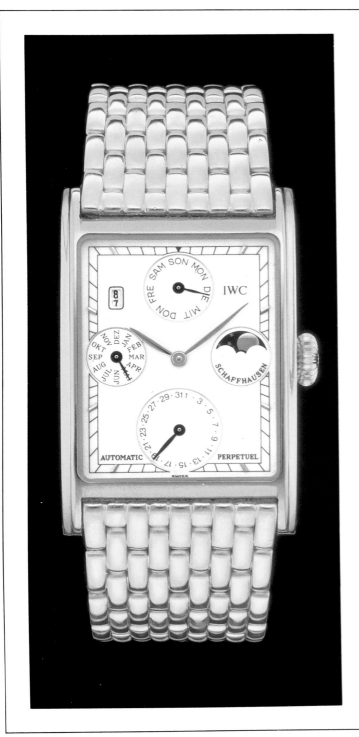

developments. This time a wristwatch was developed for military use which, according to the concepts of the contracting authorities, was to fulfill two essential requirements: the greatest possible reliability under the most demanding conditions, and protection against magnetic influences in the cockpit. IWC created it in 1940, the only firm to do so. The aviator's watch had a black dial with luminous numerals and a large center second, the movement was gilded and was set in an auxiliary inner case that was attached by three screws. The successor to this was the watertight "Mark XI" made in 1948, which was supplied to Luftwaffe units unchanged until just a few years ago. Just after the end of the war, the firm tried to reestablish old connections in Europe, and had the most success in England, where the old IWC watches had been known by the name of "Peerless".

In 1955 the owner's son took over the firm, which now did business under the name "H. E. Homberger, formerly International Watch Company".

In 1971 the firm was turned into a family stock company, and in 1978 it was taken over by the Instek AG, a branch of the German "VDO Adolf Schindling AG", and renamed "IWC—International Watch Company", its original name.

Just as varied as the firm's history has been, and just as colorful, is the assortment of

The Model 900 of 1987, with automatic winding, perpetual calendar with year and moon-phase indication; gold case and gold link band.

The famous "Ingenieur" with automatic winding, perpetual calendar, year and moon-phase indication, sweep second, case antimagnetic to 40,000 amperes/meter.

In 1985 this man's gold wristwatch, the "Da Vinci" model, came on the market, with automatic winding, perpetual calendar, year indication as well as chronograph with 30-minute and 12-hour registers. All the indications are controlled by the crown.

calibers that are produced in Schaffhausen. This development can be traced clearly, because from the first hour on, all models were numbered and registered. From 1885 on, the movement numbers were consequently recorded too, so that collectors can obtain valuable information about their watches from the books, which still exist. That is very important because Mr. Jones, for example, also had watches shipped to New York under the "Stuyvesant" name, and because the many changes of ownership resulted in "IWC" or "International Watch Company" rarely, or in some cases never, appearing on the dials.

During World War II IWC had made far-reaching developments: the factory constructed a new system for self-winding with sweep second, a similar movement with date indication, and introduced ultrasound cleaning, a method that is now common throughout the watch industry. At the beginning of the Sixties, further development of the aviator's watch led to the "Ingenieur" and finally the "Ingenieur SL", which resists magnetic fields of up to a thousand Oersteds, has a special type of shock-resistance, and is watertight to a depth of 120 meters.

In 1962 IWC recognized the beginning of the Japanese quartz offensive and took part in work at the Swiss research center for electronic developments (EH). In 1970 the firm brought one of the first Swiss quartz watches, the "Da Vinci", onto the market, using the Beta 21 caliber.

But the great upswing for IWC came in 1981, three years after VDO had taken over the firm, with the new director Günter Blümlein. At the cost of four million marks, the machine shop was moved to Vordermann. It was also Blümlein's new strategy no longer to build as many movements as possible, but rather to cut their numbers to a reasonable number, high

Man's wristwatch with automatic winding, Chronograph with 30-minute and 12-hour counters, "eternal" calendar, year and moon phase indication, "Da Vinci" model in a case of black zircon oxide, a hard ceramic material.

Small wristwatch from the "Da Vinci" line with quartz-powered watch and chronograph movements, 30-minute and 12-hour counters, date and moon phase indication, gold case, watertight to 30 meters.

Intended especially for ladies is this gold wristwatch from the "Portofino" line with quartz-powered watch and chronograph movements, 30-minute and 12-hour counters and date indication. The case is watertight to 30 meters.

118 IWC

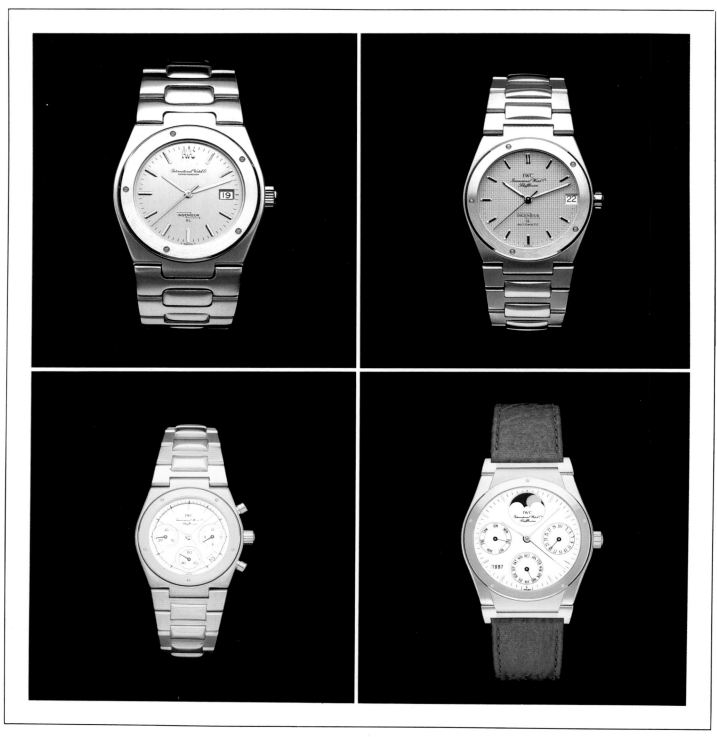

The "Ingenieur SL" with automatic winding and date indication, in its post-1976 form, steel case, antimagnetic to 40,000 amperes/meter, watertight to 120 meters.

The "Ingenieur SL" model, in present production: automatic movement with a gold rotor, date indication; case antimagnetic to 40,000 amperes/meter and watertight to 120 meters.

Likewise from the engineer line, is this gold wristwatch with chronograph, 30-minute and 12-hour counters as well as date indication; its mechanical chronograph movement is powered with the help of quartz powered step-switch motors. "Engineer" model with automatic winding, "eternal" calendar, year and moon-phase indication. The gold case is antimagnetic and watertight to 30 meters.

Man's watch from the "Engineer" line with gold case and band, automatic movement with date indicator.

IWC 119

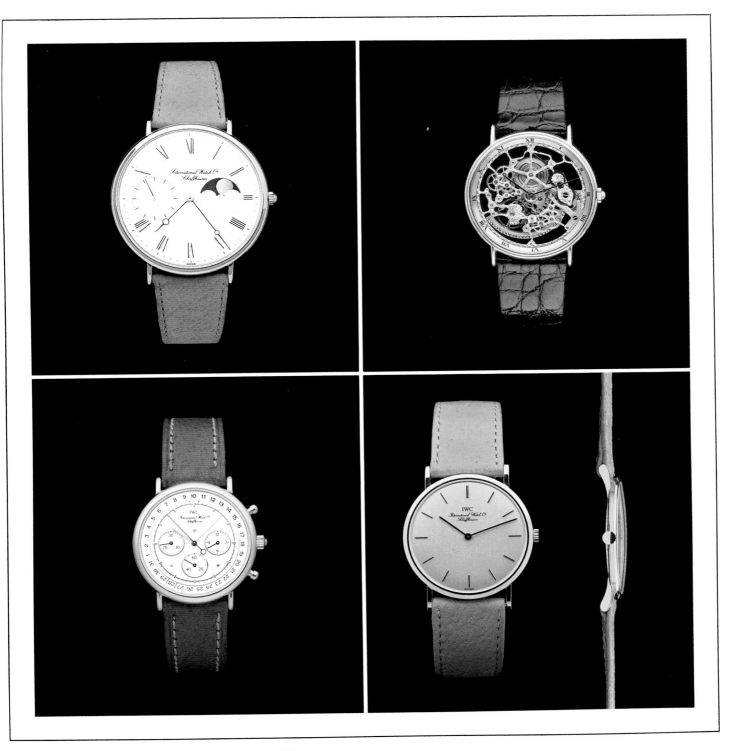

Wristwatches from the "Portofino" line. This big gold "pocket wristwatch" with moon phase indication and a pocket watch movement with hand winding was the godfather of the "Portofino" line. The movement is visible through a sapphire glass.

Man's skeleton wristwatch with automatic winding by a gold rotor.
Man's wristwatch with a quartz-powered chronograph movement, 30-minute and 12-hour counters and date indicator. The gold case is watertight to 30 meters.

Man's extra flat wrist-watch in gold equipped with either a hand-wound movement, (height 1.85 mm) or an ultra-flat quartz movement.
No longer in production is this man's gold wristwatch with automatic winding, date and moonphase indication.

Man's extra-thin gold wristwatch, with either hand winding (height 1.85 mm) or an ultra-thin quartz movement.

"Pocket wristwatch" in gold, with moon-phase indication and large hand-wound pocket watch movement, visible through a sapphire glass bottom.

120 IWC

The world's first wristwatch with a titanium case, the "Titanium Chronograph", designed by Ferdinand A. Porsche, automatic movement with day and date indication as well as chronograph with 30-minute, 12-hour registers and tachometer scale. The push-buttons are integrated ergonomically into the case rim and thus protected from damage; the case is watertight to 60 meters.

in quality, and to stress this quality instead of quantity.

His new design embodied sporting functionality with a definitely manly spirit; it caught on quickly not only in Germany, Austria, the German-speaking part of Switzerland, the Benelux states and Scandinavia (the main markets), but also in Italy, France, Spain and the United States. IWC is again out in force. With 180 employees, the firm builds and sells 20,000 watches a year. The sales figures in Germany and Italy rose 50% and 80% respectively from 1985 to 1986!

Blümlein's cooperation with Ferdinand A. Porsche, builder of the legendary "911" sports car, was not popular at first, but has proved to be a great success. One of the first results of this cooperation was the "Porsche-Design" compass watch. Today there are already five models in this line:

— the titanium "Ocean 2000" diver's watch, a civilian version of the watch that IWC built in cooperation with the German Navy.
— the "Ultra Sportivo", with quartz movement in a titanium case, watertight to a depth of 60 meters.
— the two-part "Compass Watch" with automatic winding. When the movement is raised, a 3 mm thick precision compass with shock-resistant needle appears. There is now a second type with moon phase and date.
— The automatic "Titanium Chronograph" with minute and hour registers, day and date indication. The push-buttons are integrated in the case rim so that the watch is watertight to a depth of 60 meters. It has a running duration of 48 hours.

The other watches in present production are:
— the "Yacht Club II" with a quartz movement mounted on rubber buffers, watertight to 100 meters.
— the "Da Vinci" (quartz), with optional date indication,
— the "Ingenieur SL" (quartz), in which the movement is suspended in the case to improve shock-resistance. Bottom, bezel and case are screwed; four sizes are available in gold, steel-gold and steel,
— the "Grosse Armbanduhr" (Big Wristwatch) with a pocket watch movement, moonphase indication and small second, in 18-karat gold,

"Porsche-Design" chronograph. The functions are the same as those of the "Titanium Chronograph" shown at left.

—the "Portofino" developed from it, with moon-phase and date indication and sweep second,

—an exclusive "Skeleton Wristwatch" in a gold case, and

—a number of "Ladies' Ornamental Wristwatches" with cases decorated with diamonds and other gemstones.

The top mechanical models from Schaffhausen are certainly the "Da Vinci" chronograph and the "Ingenieur" with perpetual calendar and moon-phase indication. The two models have one feature in common, namely the new perpetual calendar mechanism made just for them. The newest element of it is the digital year indication. The moon-phase indication, limited to the 29.5-day rhythm by most manufacturers, was much refined. IWC has calculated the moon-phase cycle at 29.53059 days, which differs by one day only after 122 years. In developing the movement they also were able to omit the customary corrective mechanisms. As long as the date is correct, so are all the other indications. If the watch has not been worn for some time, all the indications, as well as the time, can be set by using the crown.

It takes a year and a half to build every single "Da Vinci", which unites a number of indications on its dial:

—the exact time,
—moon-phase indication and a 30-minute register near the 12,
—the date indication by the 3,
—hour register and month indication by the 6,
—day indication and second register by the 9, and

—digital year indication in a aperature between the 7 and the 8.

The chronograph—accurate to an eighth of a second—can be started, stopped and set to zero by the two buttons beside the crown. In the 13,5-ligne automatic movement the balance

"Compass Watch", Porsche design, wristwatch with automatic winding, moon-phase and date indication; the upper part of the case rests at a 45-degree angle to allow reading of the compass underneath.

"Ocean 2000" diver's watch, developed by IWC for the German Navy. Titanium case, watertight to 2000 meters, automatic movement with date indication.

makes 28,800 beats per hour. The case is watertight to a depth of 30 meters.

The calendar is programmed to March 1, 2100. Then, of course, the 29th of February is omitted according to the rules of the Gregorian Calendar. The IWC watchmakers will have to intervene then and correct this, as well as the moon-phase indication with its built-in minimal "error" of 0.00066 days per cycle, by one day. The next programmed intervention is scheduled for December 31, 2199. Then the new century lever for the year indication, already included with the purchase, is to be put in to replace the old one, which will give the "Da Vinci" another three hundred years of life.

That the crown of the massive gold case is screwed has another reason besides that of the desired watertightness, namely that of protecting the watch from the owner's sporting activity. For the brilliantly programmed mechanical calendar can be moved forward by the winding crown, but not backward. If the owner unfortunately turns it too far ahead, then there is nothing to do but let the watch lie for the corresponding period of time or take it to a watchmaker. The "Ingenieur" has the same calendar mechanism as the "Da Vinci", but the watch does not have a chronograph.

IWC takes pride not only in these outstanding products of watchmaking art, but especially in their archives and spare-parts warehouse. In the archives, now as before, are the data of all the watches the firm has sold. In our day too, the individual movement number and caliber type, the retailer and the price the retailer paid, are still entered by hand. The initials WSCH, for example, refer to Sir Winston S. Churchill. He received his watch as a gift from eight Swiss doctors on the occasion of his speech in the Zürich University auditorium on September 19, 1946. IWC noted that the watch with movement number 955025 and case number 1107985 had left the factory on November 11, 1944. There are still replacement parts in the warehouse for watches made before the turn of the century.

Thus it can be said with pride at the House of IWC: "Our service extends from the hundred-year-old inheritance to the quartz watch with a compass."

Man's gold wristwatch with a classic IWC-automatic caliber date indication and central second. The reverse lid is fastened by a hinge, can be opened by hand, and then provides a view of the movement which is protected by a sapphire glass. The case is watertight to 30 meters.

The man's "Yacht Club II" model in steel-gold form, watertight to 100 meters, quartz movement with date indication, no longer in production.

JAEGER-LECOULTRE

A sense of perfection must have run in the LeCoultre family. At the end of the Eighteenth Century, Jacques LeCoultre lived in the isolated Vallée de Joux, about seventy kilometers from Geneva, was a well-known watchmaker and experimented with steel alloys. His goal: finding a better metal for the manufacture of watch components. This striving did not come out of thin air, for meticulous and brilliant metalworkers who passed their secrets only from father to son were always the ancestors whose history can be traced back to the Sixteenth Century in Switzerland. So Jacques LeCoultre, while he experimented with new alloys, also won a fine reputation for his work with original music boxes, keyboards and high-precision watch components.

His son, Antoine LeCoultre, who was born in 1803, took up the handworking tradition, devoted himself to watchmaking, and in 1833 founded the company from which the present-day firm has grown, which at that time was simply called "LeCoultre". Watches were the main product of the little factory in Le Sentier that stands just a few meters from the large present-day quarters and speaks of modest beginnings.

But the thirty-year-old founder of the firm very quickly showed that he was not just a gifted watchmaker, but also an imaginative fine mechanic and inventor. In 1844 he revolutionized the entire watch industry with the invention of the milliono-meter, an instrument with which, for the first time, measurements of up to one thousandth of a millimeter could be made accurately. The results for watchmaking:
—precision in the manufacture of components, and thereby in time measurement, was increased,
—the metric system became the universal measuring standard in the manufacture of watches, and brought the use of countless other systems of measurement then in use to an end.

The experts still are fascinated to see the one surviving example of the original millionometer in the firm's small museum. The solution to the riddle of how watchmakers, with the means at their disposal in the early days, produced that vital component, the ultra-exact screw, has not been preserved. It can only be presumed that his wife must also have taken an active part in its production, for Antoine LeCoultre cut this screw out of steel in their bedroom. Seven years later the invention was still so sensational that it was honored with a highly endowed gold medal at the world exhibition in London, to which LeCoultre brought pocket chronometers, pinions, wheels, and an assortment of watch movements built according to his newest developments.

The spirit of its founder (his motto "we build good watches for the future") is shown by other fine-mechanical machines,

Antoine LeCoultre

The "Memovox-Automatic", a man's wristwatch with automatic winding, date indication and alarm, from the Sixties.

such as milling machines for toothed wheels and pinions, which could already be produced in true miniature form with absolute precision in the Nineteenth Century. His lifelong maxim, "We must base our experience on science", paid off. Switzerland's oldest manufacturers recognized the precision achieved in the small watch factory, and had their own built in Le Sentier.

At that time there was not an actual LeCoultre watch, only components or raw movements, without jewels, without decoration, without balance or escapement. But these few were of such high quality that even the undisputed number one firm, Patek Philippe, made the masterpieces from the Vallée de Joux the hearts of their watches from 1847 on. Around the turn of the century there was, in fact, a period of almost ten years in which all Patek Philippe raw movements were supplied by LeCoultre. Before, as a result of a carefully preserved letter from Monsieur Philippe, the watchmakers of Le Sentier helped to build the movement production facilities of the respected Geneva firm. At first the raw movements were mounted in the Swiss Jura and finished in Geneva under the direction of the LeCoultre workshop director. Also made under contract were watches for Cartier, who bought complete products. Around 1910, when Patek Philippe was finally satisfied with the quality standards of its own manufacturing, the percentage of LeCoultre movements in their watches gradually sank to about 25%. At that time LeCoultre already produced considerable numbers. Between 1900 and 1919, 40,000 raw movements left the factory. The firm's statistics for movements with complexities showed 60,000 examples between 1860 and 1925. Deliveries were made only in quantities of at least one gross (144 pieces), at prices of 100 to 400 francs per movement.

Around the turn of the century a development occurred with which LeCoultre made the whole industry take notice: the creation of the world's thinnest watch movement, only 1.38 mm thick. The attention that the firm attracted with it encouraged them to develop the masterpiece further. Thin, thinner, extra thin, ultra-thin was the watchword that culminated in a chronograph movement (2.8 mm) and a repeat movement (3.2 mm). These were, of course, just movements, for some time was still to pass before the first complete watch left the factory. The result then was not a LeCoultre, but rather a Jaeger-LeCoultre, for the grandson of the firm's founder, David LeCoultre, "married" his company in 1925 to that of the Alsatian manufacturer Edmond Jaeger, a renowned watchmaker, supplier of Cartier and the French Navy. So the ownership did not change, but was fused. Today Jaeger LeCoultre, along with IWC, is owned by the VDO concern.

The fusion set in motion a noteworthy upswing in both technical and economic terms. In 1926 the invention of the duoplan watch was patented. The movement was designed in two planes; a balance considerably larger than was other-

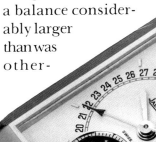

New version of a classic; rectangular wristwatch with complete simple calendar and moon-phase indication; hand-wound movement.

wise used could be housed in the lower plane, which made for more exact time measurement. In the same year the first case of stainless steel was produced. The way to their own watch, two years later, had been paved.

In 1925 another "world's first" emerged from the factory: Jaeger-LeCoultre offered the world's first wristwatch with a scratch-resistant (supplied from France) sapphire crystal. The other manufacturers of high-quality watches followed suit, and from then on the broken or scratched crystal of a watch in this price class was a thing of the past. In the same year another bold venture succeeded, in the form of the smallest mechanical watch in the world. The movement consists of 74 components, which are contained in only 4.85 x 14 millimeters. The weight of the movement with its fifteen ruby jewels adds up, including the dial, to less than one gram. A few years after its development, record production amounted to ten pieces per month.

With this watch the factory put itself at the top, where it remains today, unequaled. The most prominent owner of one of these watches, of which some thirty are still made per year, is Queen Elizabeth II. Her ornate specimen is made of white gold, with diamonds on the case and band. The value of the movement alone is estimated at about 3000 Swiss francs.

In 1931 the idea that "the crystal must be protected" was turned to reality in a truly brilliant, simple way. The company created the "Reverso", a successful model that, after a forced pause during the war, was put back into production by chance and has now risen to the top with 2500 pieces produced per year (about 12,000 in all) in terms of finished watches (not to be confused with the production of movements, which number about 100,000 per year).

But to take things in order: the best way to protect the crystal, thought the Jaeger-LeCoultre watchmakers, would be for the watch not to have a crystal. It only needs one a few moments a day, namely when the owner wants to see what time it is. So the watch case was set on a massive bottom plate linked with the band, made so that the case and thus the watch can be turned 180 degrees, soundless, precise. The crystal thus disappeared into the interior of the band, and a decorative plate presented itself to the observer. The slogan for this watch: "The Reverso is a piece of jewelry on the one hand, and a watch on the other. And vice versa..."

It was a sensation on the market in 1931, for at the time only a few elegant wristwatches were robust enough to withstand the strains of a polo match at Windsor, a skiing trip in the high Alps, or a thousand-mile auto race in Sicily. An Italian retailer deserves thanks for the fact that the watch "with the twist" is being made again today in almost unchanged form. For during the war this line was no longer in production, and as a result it had been forgotten. At the end of the Sixties, that Italian dealer was visiting Le Sentier to see the new collection. While being shown around the factory, he hap-

Man's wristwatch with complete simple calendar and moon-phase indication. The original dates from the second half of the Forties. A new series was begun in 1982 using original parts.

Woman's ornamental watch with gold band, attachments and bezel decorated with diamonds and rubies.

Three models from the "Odysseus" line first offered in 1988, from left to right.

Man's gold wristwatch with mechanical chronograph movement of 24 mm diameter and height of 3.7 mm. Watch and chronograph movements each powered by a quartz step-switch motor. The watch also has 30-minute and 12-hour counters, date, and moon phase indication.

Man's gold wristwatch with automatic winding, "eternal" calendar, year and moon phase indication. In the middle of the dial is an auxilliary 24-hour indication.

128 JAEGER-LECOULTRE

Various
"Reverso" models
in gold.

These gold models of the "Odysseus" line with automatic or quartz movement were introduced in 1988.

pened to look in a few drawers and suddenly pulled out a few old Reverso cases. He wanted to know what they were. And the watchmaker at whose workbench this happened said that they used to make such a watch. That was just old junk, really, that for some no longer justifiable reasons he didn't want to throw away. The Italian thought he might be able to use the old junk, so he took the last six cases with him. Back in Italy he built a suitable caliber in, put the watches in his shop window and sold them in a few days.

He needed more of them, so he telephoned to Le Sentier, but Jaeger-LeCoultre had to decline. "We don't have any more", was the laconic answer. But when the Italian reported a few days later that his customers, who had seen the watches on others, were simply wild about them and wanted one for themselves, he set off a spark in Switzerland. In 1979 the "Reverso" went back into production. The watch, which became a classic by chance, is now available in three sizes, with different dials, with quartz movement or, in a few individual cases, with mechanical movement, with leather or massive gold band, with or without diamonds, in steel-gold and, since 1987, also in stainless steel. The reason for the last version: "When one does not know which way the economic development is going, the trend is to the steel watch. And that is how people reacted." A monument really ought to be put up at Rue de la Golisse 8 to the Italian who discovered the old cases, for—as noted— today the "Reverso" is the favorite among the masters of Jaeger-LeCoultre, who, according to a tenet of the firm, look for a full harmony between caliber (thus tech-

In 1986, for the 35th birthday of the "Memovox", a series of wristwatches limited to 350 pieces and individually numbered (see engraving on the bottom), newly redesigned with automatic winding, alarm and date indication.

nology) and design in their watches.

In technical terms, the factory has offered additional masterpieces since the war as proof of their ability:

— In 1953 the watchmakers brought out the first fully automatic watch, the "Futurematic". Whereas at that time a conventional automatic wristwatch that had not been worn for a long time had to be started by manual winding, the new development no longer has any winding mechanism. Instead it has a "maximized" running duration and—as a "by-product", since there was space available in the movement—an extra-large balance, contributing to its precision. There was a small winding indicator on the dial to let the owner read the amount of reserve. And if the watch were to stop, a light shaking was enough to wind up the movement again sufficiently.

— In 1967 the factory produced an automatic movement only 2.35 mm thick.

— In 1979, the same year in which the classic "Reverso" was reborn, a 2 mm thick quartz movement for women's watches was put on the market.

Since then this movement has been reduced to a thickness of 1.8 mm (of which the battery alone requires 1.1 mm), and 9.5 x 11.5 mm dimensions (including the battery, which has a diameter of 6.6 mm and, thanks to a particularly power-sparing switching, only needs to be changed after five years).

For more than 150 years masterpieces keep coming from the factory in Le Sentier, as complete watches that are sold worldwide by more than 2000 retailers, or as raw movements that are still delivered today (in part exclusively) to the finest addresses in the Swiss watch industry:

— Vacheron Constantin,
— Audemars Piguet,
— Piaget (the world's smallest quartz movement),
— Baume & Mercier (three different calibers),
— Chopard,
— IWC, and
— Corum.

Jaeger-LeCoultre equips 85% of their own watches with quartz movements at this time. Women's watches add up to 40% of total production. Most of the watches are sold in Switzerland, the rest go primarily to Europe, the USA, and Japan. White-gold models have been added to the program especially for Japan because, the sales director explains, in that country white gold is not regarded as ostentatious, as is yellow gold.

It is also noteworthy that Jaeger-LeCoultre has been successful in recent years not only

Models from the "Albatross" line in steel-gold, watertight to 100 meters. At left with automatic movement, complete simple calendar and moon-phase indication. At right with date indication.

with the "Reverso" but also with other replicas: In 1933 two old moon-phase watches were made in limited numbers for the firm's anniversary. The "round moonphase" of 1946 (250 pieces) with identical movement and the "rectangular moonphase watch" (some in new form) of 1940 (250 pieces), in which a new type of mechanism switches the month indication.

In 1986 the "Memovox" was redesigned (350 pieces); it had been fitted out as one of the first alarm wristwatches in 1951, equipped with two barrels, one each for wheel train and alarm, and has caused a sensation. It had two crowns, one for winding and hand-setting, the other to set the alarm, whereby the alarm crown had to be pulled to set the alarm, and pushing it back in wound the alarm, which always "buzzed" mechanically for twenty seconds. Thanks to quartz technology, alarm watches were nothing new any more, but the "Memovox" remained special, as in its new form it was the only watch with mechanical alarm and automatic winding.

Today the classic "Reverso" is the popularity leader among Jaeger-LeCoultre watches, followed by the "Albatross" collection. The Albatross is referred to as the "genuine factory product". This is justified, as all of it—from case to movement to band—is produced in the house, thanks to robots which eliminate mechanical work in part, but mainly made by high-quality handwork. 150 individual parts have to be assembled to make the gold or steel-gold band alone. The hexagonal dial and the thin profile are typical of the "Albatross". The name has changed during its lifetime: a Roman II has been added, for the "Albatross II", a sporting-elegant watch watertight to 100 meters now has, by popular demand, a date or moon-phase indication and/or a bezel set with diamonds. Quartz movements are usually used, but automatic mechanical movements are also available on request.

Third place on the Jaeger-LeCoultre popularity scale is taken by the "Lyre". This classic round watch is made in three sizes. In 1986 some new "Lyre" models were created, which are watertight to 30 meters with one exception, a model with 107 diamonds set on its gold case. The other "Lyre" models have 54 (man's watch) or 46 (woman's watch) diamonds on their bezels. Two-color (steel-gold) cases are also new, as are models with moon phase and/or date. The "Lyre" with digital year indication was added in 1987.

A moon-phase watch with automatic winding, complete date indication and transparent bottom is made only as a man's watch. The ornamental "Rivière" and "Semi-Rivière" watches are made of yellow or white gold, and they are equipped with either the smallest mechanical movement in the world or the quartz movement with the smallest volume. In the spring of 1987, Jaeger-Le-

Man's gold wristwatch in the form of a padlock, hand-wound movement; mid-Seventies.

Coultre presented the new "Vogue" line. Here the firm uses exclusively mechanical movements with small second and classic design. The watches are made and sold in gold and steel-gold. In addition, a watch with automatic winding, perpetual calendar and year indication was exhibited at the 1987 exposition in Basel. The calendar mechanism comes from the "sister" firm, IWC, the movement from the house itself.

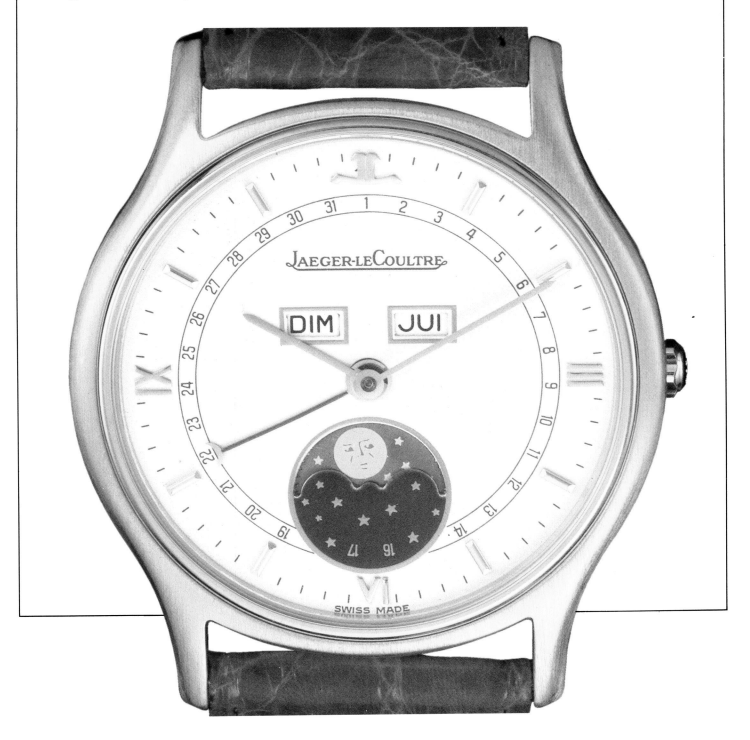

Man's gold wristwatch with automatic movement, date, day, month and moon-phase indication.

134 JAEGER-LECOULTRE

Even though we are talking of wristwatches here, there are two specialties of the Jaeger-LeCoultre factory that deserve mention, as they offer insight into the firm's philosophy.

— The know-how of the fine mechanics and jewelers is marketed in other ways. So the company has another leg to stand on— in the manufacture of exclusive lighters and ball-point pens (under the Christian Dior brand name).

— The second specialty is the unique "Atmos" clock.

The principle was discovered in 1928 by the Neuenburg engineer Jean-Léon Reutter. Jaeger-LeCoultre developed it further for series production. In a capsule is a (secret) liquid and a (likewise secret) gas. The mixture either expands or contracts (according to the temperature). This activates a bellows like an accordion in the capsule, which winds the mainspring of the movement. The required energy is minimal, of interest only to a mathematician; sixty million of these clocks use no more energy than a 15-watt light bulb. Since all parts operate almost contact-free, Jaeger-LeCoultre states a life expectancy of about six hundred years for this watch. With a pessimistic note, to be sure: "With today's air pollution, we must unfortunately advise you to have the

The models of the "Gaia" line, first exhibited in 1987, are reminiscent of the early years of wristwatches. The winding crown is located near the 12. The men's models are hand-wound, the women's have quartz movements.

The "Reverso" for women, with a gold link band and quartz movement.

watch cleaned every twenty years".

The "Presidential Atmos" were given by the Swiss government to John F. Kennedy, Sir Winston Churchill, General Charles de Gaulle, Habib Bourguiba, Princess Irene of The Netherlands, Giovanni Gronchi, James McDivitt, Edgar White, Richard Nixon, Nelson Rockefeller, King Hussein of Jordan, Prime Minister Harold Wilson, Lee Kuan Yew, Pope John Paul II, Charlie Chaplin, Ronald Reagan and Mikhail Gorbachev. They are recorded as owners in the firm's "Golden book".

A gas-powered wristwatch was developed as early as 1945 but never put into series production.

At Jaeger-LeCoultre too, wristwatches are entered in a register upon delivery. Along with the reference number, the detailist is recorded; he notes the case number, and with his help the factory can give information at any time as to the history of the watch. There are no problems with written inquiries, as the detailist considers himself responsible to "the large family". Repairing of watches made since 1935 is just as unproblematic. Generally spare parts are available, and if not, they can be made separately, as with older watches. A watch from the Thirties, with broken crystal and rusted dial, was just being repaired when we visited Le Sentier. The crystal was still available, the dial was made new. It was not cheap (in this case, over a thousand francs), but the customer always receives an estimate in advance.

The owner of a Jaeger-LeCoultre watch can thus rely on one thing: the manufacturers will always try to find an acceptable solution to all problems. For they see the customer too as a member of the big family.

One of the new creations in the "Reverso" line: the model with additional moon-phase indication.

The "Atmos", the clock that lives on air. The winding of the clock is done by temperature variations in a capsule.

PATEK PHILIPPE

Tsar Nicholas I, who ruled the Russian Empire from 1825 to 1855, wanted to change the world with his Polish policy. In passing, he did. But he probably never thought while looking at his opulent pocket watches that he would activate the foundation of one of the world's outstanding watchmaking companies in faraway Switzerland.

But the fact is that after the Congress of Vienna nothing remained of the Polish kingdom but the so-called "Congress Poland", which was united with Russia and yet had a constitution that guaranteed its national

integrity. In the little town of Piaski in this "rump Poland", completely ignored by historians, lived Joachim and Anna Patek de Prawdzic. On June 12, 1812 a son and heir, whom they named Antoni Norbert, was born to them. The life of this Antoni Norbert Patek de Prawdzic, who enlisted in the First Protection Company of the Polish cavalry, was to be fully changed by the Russian Tsar's decisions and lead to the founding of the noble Swiss brand of Patek Philippe.

The Polish army in which Antoni Norbert served in 1830 was under the command of the Tsar's brother, Grand Duke Constantine. Tsar Nicholas relied so much on "his" Poland that he even wanted to send part of its army to Belgium and France to support the revolution there. The intentions of the Tsar and his brother became public through poor security and led to the fact that even before the order to march westward was given, a group of cadets moved eastward and started a revolt in Warsaw. They occupied the Belvedere Palace, executed two Russian generals, and inspired Grand Duke Constantine to flee. When the Polish parliament decided on January 18, 1831, no longer to recognize the rule of the Tsar and, anticipating the results of this bold step, increased the army from 30,000 to 80,000 men, Tsar Nicholas I sent Russian troops marching into Poland. There were vigorous battles, in which Antoni Norbert Patek went into the field as a freedom fighter alongside his countrymen, was wounded twice, and was decorated with the golden cross of "Virtuti Militari" after his promotion to second lieutenant. But at the start of 1832, the Polish revolution was put down and a ruthless policy of russification was introduced. Thousands of Poles and large portions of their army fled from the Tsar's thirst for revenge and orders to have rebels shot.

At first Patek found political asylum in Paris, but was ordered to Bamberg by a Polish general, who wanted to organize the expatriate

Left: Antoine Norbert de Patek (1812-1877).

Right: Jean Adrien Philippe (1815-1894).

army members, to serve as a representative and liaison officer to the refugee committee. When he had fulfilled his duties to the satisfaction of everyone involved, he moved back to France. He worked as a typesetter in printing shops in Cahors and Amiens, and finally went to Geneva, the center of watchmaking, jewelry production and artistic handwork. At this time he was not thinking about watches. Thinking in terms of adapting his name to the linguistic customs of his new homeland, he called himself Antoine Norbert de Patek from then on.

Soon he discovered a tendency toward the artistic in himself and began to study painting with the famous landscape painter Alexandre Calame. Not much of this period has been recorded, but it is known that his studies also included a stay of several months in Paris, at the end of which Antoine Norbert de Patek got into the watch business. He bought particularly good movements from masters of the art and had them built under his supervision into cases that were masterfully made and decorated by goldsmiths, engravers, chasers, enamelers and miniature painters, making them into very artistic pocket watches. About twenty-five years old then, he put special emphasis on the highest quality in both the technical and the artistic aspects in the producing of "his" watches.

But in these times he had not forgotten his countrymen. With several exiled Poles, he made efforts to found the "Polish Foundation", which was actually organized a year later, in 1839. Antoine Norbert de Patek assumed the position of treasurer.

But this year of 1839 had even more decisive events in store for the now-successful watch "dealer":
—He married Marie Adelaide Elisabeth Thomasine Dénizart, the daughter of a French merchant, in Versoix, near Geneva.
—He met François Czapek and founded with him the firm of Patek, Czapek & Co.

Thus he had an ideal watchmaker in the business, for François Czapek had learned this trade from the ground up. Past experiences probably had an effect on the partnership, for the Polish-Bohemian, who was scarcely a year younger than his business partner, had also taken part in the Polish revolt as a soldier in the National Guard, and only afterward learned the watchmaking trade in Vienna and Prague. Then he too had been drawn to the Mecca of watchmaking. Even before their partnership was formed, the two recognized that with the heavy competition in the hotly contested market within the city walls of Geneva, the only hope of mercantile success existed for those who were not satisfied with mediocre goods but strove for especially high-quality results.

This became their guiding principle as they worked with five to seven colleagues in their workshop at Quai des Bergues 29, making about two hundred

The firm's headquarters on the Grand Quai of Geneva from 1854 to 1890.

pocket watches a year, some with repeat striking that stood out technically as well as artistically by its first-class work. The examples still in existence today give evidence of the quality of their masterpieces. But these pocket watches also show, by their many Polish-language engravings and the choice of artistic motifs that it was mainly Polish exiles who had them made by their emigrated countrymen. Pictures of the Madonna of Czestochowa and the Madonna of Ostrobrama had to be painted on the lids of the pocket watches particularly often.

From the first watch on that originated in the small partnership, a consecutive number was punched into the bottom plate of the movement, but only from number 63 on were the data on them recorded in a book. From these handwritten notes it can be learned that Patek & Czapek, on November 21, 1839, produced the first pocket watch with the crown winding made by the watch technician Louis Audemars, of the Vallée de Joux. This system allowed the watch to be wound without using the key that was still customary then, and that often got lost and had to be replaced—still the lesser evil compared to the price of a complete watch.

In 1843 Antoine Norbert de Patek became a Swiss citizen, but still felt linked with his former countrymen, and in 1844 he took part in the establishment of a Polish library and reading room in Geneva. While he openly struggled to overcome his homesickness in this way, his partnership with Czapek was more likely to assuage his feelings for his homeland through extended trips to Poland and Bohemia. There were more and more differences of opinion,

Man's gold wristwatch with enamel dial, from the Twenties.

Man's cushion-shaped wristwatch with chronograph, 30-minute register and tachometer scale; circa 1930.

Above: men's wristwatches with automatic winding, lever movement with machined gold rotor, circa 1954.

Below: Men's wristwatches in which the hour hands, or the blue hour hand, can be set by the button located on the left side of the case rim, independently of the minute hand. In this way the time in another time zone can easily be set. Watch movements with hand winding. The model at left was made around 1960, that at right from 1962 on.

Man's wristwatch with chronograph, 30-minute register, perpetual calendar and moon-phase indication; model 1518, produced from 1941 to 1954.

Man's wristwatch with chronograph, 30-minute register, perpetual calendar and moon-phase indication, model 2499, produced from 1950 to 1985, at first with rectangular, later with round buttons to operate the chronograph mechanism.

Man's wristwatch with chronograph and 30-minute counter, model 1579, produced from 1943 on in yellow gold, red gold, platinum, steel and steel-gold cases.

PATEK PHILIPPE 141

Patek traveled to Paris, where he could exhibit his watches to a lot of people at a big exhibition for the first time. This exhibition was to become a milestone in his life! While making the rounds, he met the three-year younger (29 at that time) Parisian watchmaker Jean Adrien Philippe, who was attracting attention with his particularly thin pocket watches.

Jean Adrien Philippe had inherited the love for mechanics from his father, who had worked as a small watchmaker in the village of La Bazoches-Gouét, some ninety kilometers southwest of Paris, where, fully unknown to the great masters of his time, he had made pocket watches with very original mechanisms for striking, calendar and moon-phase indications. He had not gotten rich from it, but he had awakened an intense interest in this kind of handwork in his son. From childhood on, Jean-Antoine had sat, first on the workbench, and later, when he was too big for that, next to it, watched his father, asked many questions, and tried working with the tools and materials himself. At eighteen he came to the conclusion that his father could not teach him anything else. He walked to Le Havre and worked with a chronometer maker, who lent him a few books that the young mechanic often studied till late at night in order to learn as much as possible.

Along with Switzerland, England was then a center of the high art of watchmaking, and in 1836 the eager young handworker sailed to London to complete his training. Not much is known of his time in the English capital. It is certain, though, that Jean Adrien Philippe

and often loud quarrels between the two, for Patek was concerned about the continuing development of the business, which in any case was faced with either an extension or a dissolution of the partnership.

In 1844 the busineslike Antoine Norbert

Cushion-shaped gold wrist-watch made in 1908.

Men's rectangular wristwatches:
Left: made since 1948.
Center: 1945.
Right: made since 1948.

Man's rectangular wristwatch from the first half of the Twenties.

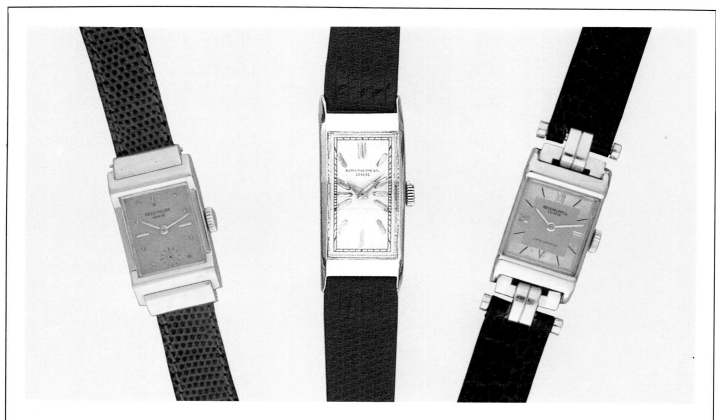

and another watchmaker, whose name is not known but whom he met in England, finally went to Paris. The two young men had an ambitious goal: they wanted to build the best watches in the French capital. This was probably not as hard as it seems at first glance, for Paris had long since given up the title of "watch capital" to Geneva, and still stayed in the field only through the fame of a few old masters. A member of the government of the "Citizen King", Louis Philippe of Orléans, must have seen this too, for a goodly sum of money soon flowed out of the state treasury into a new business that had been founded in Versailles: a watch factory. With a lot of enthusiasm and tireless personal dedication, a yearly production of 150 watches was achieved in a remarkably short time. This was really a huge number, because except for the cylinder escapements, every single part was made in the little factory, from toothed wheels to blocks, from dials and hands to cases, piece by piece.

Since the state, as is well known, would rather take than give, financial shortages were all but inevitable, for the loans had to be paid back on time. In 1842 Adrien Philippe saw only one way out of the crisis: the factory had to bring a unique pocket watch onto the market. Since at that moment the eternal dictates of fashion called for thin pocket watches, he looked for ways to make his own

Left and right: men's rectangular wristwatches in imaginative cases; Forties.

Center: man's rectangular wristwatch from the Forties, platinum case.

PATEK PHILIPPE 143

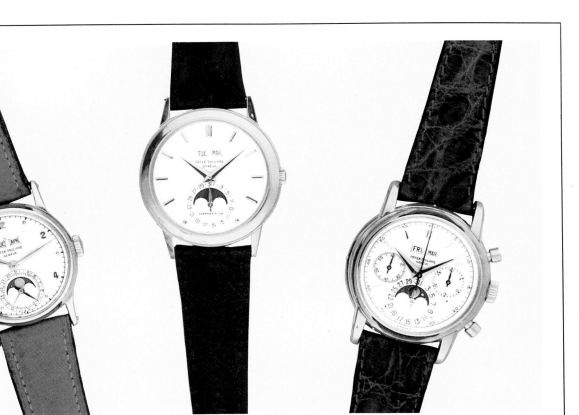

products fit into this trend. The goal, he found, had to be reached through a new kind of crown winding. A case maker had once explained to him in general terms the system of the House of Breguet, and a watchmaker had explained the function of crown winding after the system of Louis Audemars, which Patek and Czapek also used. But in the Breguet system Adrien Philippe saw the basis of developing thin pocket watches without keys.

He worked doggedly, designed, constructed, discarded many solutions, finally succeeded and gained no recognition from the Paris watchmakers to whom

Men's wristwatches with perpetual calendar and moon-phase indication:
Left: with hand winding, made from 1942 to 1952.
Center: with automatic winding, made from 1962 to 1982.
Right, with hand winding and chronograph, made from 1950 to 1985.
Men's large wristwatch with chronograph, 30-minute register and tachometer scale: 1952.

Man's wristwatch with chronograph and 30-minute register, from the Forties: dial with an additional tachometer scale in miles.

Man's gold watch with movable band attachments. The Lépine movement from the time around 1902 may well have been built into this case in the Thirties; noteworthy is the winding crown by the 12.

144 PATEK PHILIPPE

Man's barrel-shaped wristwatch with minute repeat, platinum case (40.8 grams), sold in 1929 to H. Graves Jr. for 7700 swiss francs.

View of the 11 ligne bridge movement.

Back of the case with engraving (arms of the Graves family).

Under-the-dial view (repeat striking mechanism).

PATEK PHILIPPE 145

From upper left to lower right:

Model made from 1937 on.

Wristwatch with world time indication, made from 1937 on.

Man's watch from the Thirties.

Model from the second half of the Thirties.

Man's watch from the mid-Thirties.

Octagonal wristwatch, beginning of the Thirties.

Man's rectangular wristwatch from the first half of the Thirties. Many-faceted crystal and upper side of the case.

Man's watch with complete simple calendar in digital form; 1941.

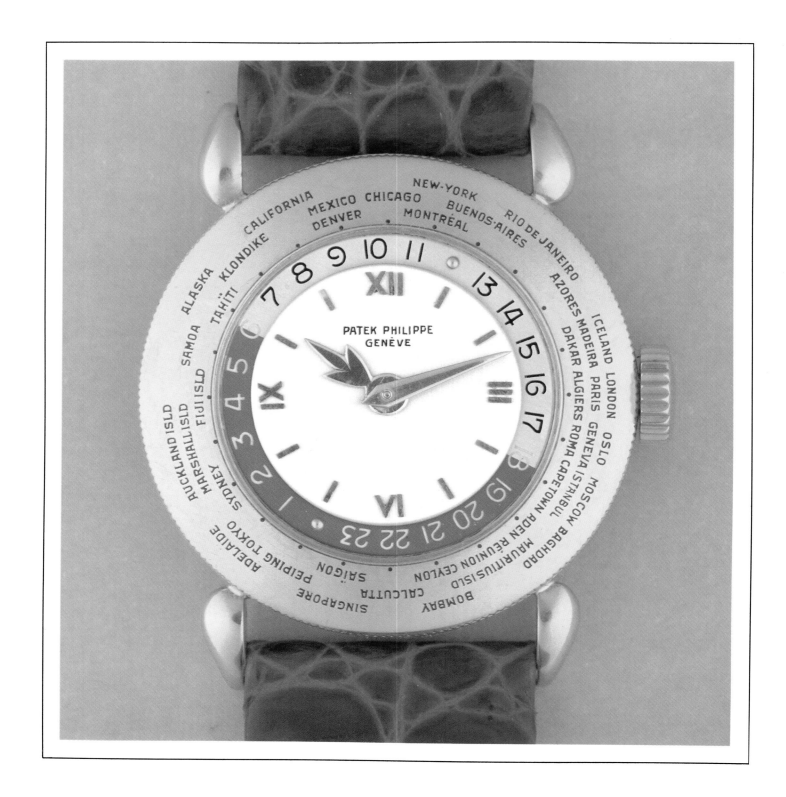

Man's wristwatch with world time indication; model 1415, made from 1939 on in yellow and red gold cases.

Man's wristwatch, with dial formed in Cloisonne enamel style. Hand-wound movement with sweep second. Model 2481, made since 1950.

Men's gold wristwatches with automatic winding, perpetual calendar and moon-phase indication, made from 1962 to 1982.

Men's gold wristwatches with perpetual calendar and moon-phase indication:

Left: made from 1952 to 1963.

Right: made since 1951.

Lower right: model with retrograde date indication; 1937.

he offered his new products. His reward was rejection and disappointment. The aris exhibition of 1844 seemed to be his last hope of making a breakthrough with his flat pocket watches with crown winding.

He was doubly rewarded:
—He was honored with a gold medal for his extra-thin pocket watches, and
—he met Antoine Norbert Patek, who was on a search for new products.

The latter already had been able to gain many years' experience with the other system of crown winding and assured Adrien Philippe that his pocket watches had a great future. The two watch enthusiasts talked with each other for hours. After that, Antoine Norbert de Patek was so impressed that he decided not to extend his contract with Czapek, but to found a new company as soon as Philippe was ready to work with him as his technical director. Adrien Philippe worked out a deal and agreed.

On May 15, 1845, the new firm of Patek & Co. was founded at Quai des Bergues 15 in Geneva, only seven houses from the former firm of Patek & Czapek. The third member at first was the judge and attorney Vincent Gostkowski, likewise a Polish emigrant living in Geneva. Patek's original partner, François Czapek, continued in business alone. For a few years he was allowed, for unknown reasons, to make use of the new crown winding in his own products without paying license license fees. Czapek was still to found branches in Poland and Paris, but finally dissolved the firm in 1869, after he had, in 1850, written the first Polish book on watch technology: "A Few Words about the Watchmaker's Trade for the Use of the Watchmaker and the Public".

Man's wristwatch with hand-wound movement, porcelain-white dial, bezel with hammered nail pattern, present production.

The firm of Patek & Co., whose forerunner had produced almost nothing but individual pieces because they were finished by hand and thus always different, was founded at a time when the Swiss watch business was changing. It stood on the threshold of industrialization, at the beginning of the road to series production of watches. Almost all the Swiss watchmakers, unlike their English colleagues, had recognized this. As a result, the Swiss could steadily increase their share of the world market, while English watch production gradually lost its importance.

As a result of this trend, Adrien Philippe had to devote his constructive skills not only to developing crown winding further, but also, in the very first year of the new firm, to creating a series of machines with which the production of watches could be done effectively and with high quality. The firm's products had not essentially changed at first, and a large number of clients were still Polish emigrants, so that the same motifs were seen on the case lids as before. But now most watches had Adrien Philippe's crown winding, which was first patented in 1845. Business went remarkably well. Remarkably indeed, in that in 1848-49 France was seized by a revolutionary movement that spread to Germany, Austria, Hungary and Italy. In this time of many crises, Patek & Co. built more than four hundred watches per year, exactly 2618 between 1845 and 1850.

To evaluate the factory's production capability realistically, one must realize that the firm of Patek & Co. was one of the few companies that could keep their watchmakers working full-time. In fact, the situation in the watch business in Geneva at that time was so precarious that the city government even instituted a make-work program for trained watchmakers. In those days one saw men who

Men's gold wristwatches from the Seventies. The second model from left (1970) is fitted with the quartz caliber Beta 21.

knew how to do excellent work with the smallest tools working with picks and shovels at a hastily created building site. They tore down the medieval city walls and transported tons of heavy stones and rubble out of the city. It was no wonder that at that time a job at Patek & Co. was simply the best thing that could happen to a Geneva watchmaker. That was surely one reason why the young company was able to use its highly qualified work force to dominate the market with its products.

On the other hand, the mercantile talents of Antoine Norbert de Patek were also responsible for the steady success. In 1848 he began to make long business trips for the purpose of bringing his watches to the people, for in those days the customers avoided long trips whenever possible. So it was probably this "marketing" as well that assured sales in Europe and America. The artistically talented merchant even took a sea voyage across the Atlantic Ocean. He maintained contact with Geneva with countless letters which are preserved in the firm's museum today and give information about the dangers and difficulties of these business trips.

At this time the production of the world's smallest pocket watch also took place; a year later it was to cause a sensation at the World's Fair in London. It can be regarded, on account of its size, as a forerunner of the wristwatch. This "little watch" with a movement of barely four lignes, had a wild rose in enamelwork on its case lid. It was sold to London in 1851 for 3750 francs, came back to Geneva in 1853, and changed owners again in 1855 for 5600 francs.

In 1851 the three partners decided to rename the firm Patek Philippe & Co., in order to document publicly the noteworthy achievements of Adrien Philippe. In the same year the "new" company achieved great international recognition in the trade. This was gained at

Wristwatches from the Seventies with various dials and decorations.

the aforementioned first World's Fair in the Crystal Palace, newly erected in London's Hyde Park. In the official catalog, number 274 (Patek Philippe & Co., formerly Patek & Co., Geneva, manufacturers and inventors) took up more space than any other Swiss watch manufacturer present. The text, with many illustrations, probably gives the best overview of the watches then produced in Geneva:

— "Complete assortment of high-quality pocket watches with all modern improvements and the most varied decorations: simple pocket watches, those with repeat, with self-acting striking and with touch hands for the blind; pocket watches with independently springing second hands and date indication, also with protected marine compass, small telescope and secret compartment; likewise pocket watches with the designation 'à triple effet', which can be changed into three different outward appearances."

— "The smallest pocket watch ever produced, the movement of which has a diameter of no more than three and three-quarter lignes."

— "The collection includes simple pocket chronometers as well as those with repeat mechanisms, tested by astronomical observatories and supplied with official running certification."

— "The winding and hand-setting of most of the pocket watches exhibited are accomplished without a key, with a mechanism invented by the exhibitor, so simple and robust that is can be used in any pocket watch, even those with two barrels and independent springing second hand or automatic striking. In addition to great convenience, this invention brings with it the advantage of not having to open the watch to wind it. This prevents the entry of dust and moisture into the movement, which results in longer usefulness of the oil."

— "Models of not yet gilded watch movements, which are intended to show the new types of production methods, and movement components that are produced by machines and tools invented by the exhibitor."

— "The raw movements of the pocket watches were both designed and manufactured by the exhibitor, as well as all other parts of the simplest to the most complex pocket watches and chronometers assembled in his factories, including the machining, engraving and chasing, the setting of precious stones and the enamel painting of flowers, landscapes, portraits and historical subjects."

Wristwatches from the Seventies.

Man's gold wristwatch with chronograph, 30-minute register, perpetual calendar, moon-phase, leap-year and 24-hour indication; first public showing in 1986.

For Patek Philippe the high point of the fair was the visit of the British royal family. Queen Victoria and Prince Albert were so enthusiastic about the displayed pieces that they bought two watches on the spot: Queen Victoria bought watch number 4536, which is presently to be seen in the firm's museum. This woman's pocket watch hangs from a blue-enameled brooch of 18-karat gold, decorated with thirteen diamonds. The cover of the gold case is likewise enameled in blue, ornamented with engraved flowers, and set in the center with diamond roses (twenty-four stones). In the case is a twelve-ligne movement (barely 27 mm in diameter) with cylinder escapement, ten jewels and crown winding. Prince Albert purchased a heavy gold hunting cased watch (number 3218, made around the beginning of 1850, thus still under the name of Patek & Co.) with chronometer escapement, quarter-hour repeat striking and crown winding.

Their success in London brought to Patek Philippe & Co. the awareness that world expositions were the best form of advertising their products. As a result, special watches were always produced for such events. It also resulted that the demand for their creations grew steadily, and thus the rented quarters at Quai de Bergues 15 grew too small. Patek Philippe looked for a new headquarters for the firm, and in 1854 rented a new building on the Grand Quai, right on the shore of the Lake of Geneva, between two hotels, which has been the seat of the company ever since. At that time ships docked right in front of the house, and the barometer column, which now stands on the Quai Pierre Fatio, stood right in front of the entrance to the building, on the ground floor of which comfortable sales rooms were located. Production took place in the upper stories.

Shortly after moving, Antoine Norbert de Patek made another long trip, particularly to work for greater sales of his watches on the American market. The situation in North America, in terms of domestic policy, was typified by the fight against slavery. The trip was not always without danger, as Patek's writings show. In June of the following year, 1855, he was on the continent of Europe again, just at the right time to participate in the Paris World's Fair. To describe the strains of such business trips, we can quote from

Woman's and man's wristwatch with gold band; present "Calatrava" line.

Antoine Norbert de Patek's numerous letters.

On November 23, 1854, the "traveler on the subject of watches" first arrived in the capital of France. He was not completely happy about his self-imposed mission, for he wrote to Geneva, among other things:

"My friends, the difficulties of the trip are beginning now. When will we be able to sell watches favorably and then wait for the customers at home, instead of having to travel all over the world with our products, incurring high costs and endangering our health?"

He had enough time to talk business with the watch dealers in Paris, for the ship "Niagara", on which he originally intended to travel farther, first had to transport troops to the east during the Crimean War (the English and French were then on the side of the Turks against the Rusians). What his Parisian colleagues told him did not exactly sound encouraging. Antoine Norbert de Patek wrote:

"I have talked with all the watchmakers known to us. They all complain of the slackness of business. And yet—business goes on."

The trip went on too. On November 29 the Swiss left the harbor of Liverpool on the "Pacific". On December 14 the paddle steamer arrived in New York. One day later the traveler again took up pen and paper:

Men's gold wristwatches, end of the Seventies.

"Life in New York is very expensive. Since I am staying in first-class hotels for reasons of security, I cannot get along on less than 24 to 25 Swiss francs a day. Room and board cost two and a half dollars a day, a small bottle of wine costs a dollar, and the cheapest cigars cost four cents." ... Mr. Reed and Mr. Tiffany (Tiffany had already been a client of Patek's since the Forties) were very surprised to see me. On account of the general crisis situation they could not promise to buy anything. Their firm is simply colossal. They have a stock valued at six million Swiss francs."

... "New York, December 22, 1854. In America as in Europe, the people are of the opinion that the coming year will be even more disastrous for business because of possible disruptions in Europe. Austria and Prussia could likewise be drawn in. (A fear that concerned the Crimean War but did not come true.) ... "If I should not be able to sell the pocket watch with the portrait of Jenny Lind (the soprano celebrated in Europe as the "Swedish Nightingale"), I'll bring it back to Europe and perhaps display it in Paris."

... "New York, December 26, 1854. I shall do my best to advertise our products, but it doesn't help much here. The recommendations of the press are not always followed, and the American trusts only what he can see."

Men's gold wristwatches from the Forties.

Man's gold wristwatch with minute repeat, perpetual calendar, moonphase indication as well as chronograph with 30-minute register; individual piece.

Men's and women's watches with gold bands, present production.

..."Boston, January 9, 1855. We must display in Paris and win a medal. I am happy to be able to tell you that we were honored with the silver medal at the international exposition in New York."

The trip continued, from Boston via Philadelphia and Washington to Charleston, where Antoine Norbert de Patek put the following on paper on January 26: "The Americans want not-too-expensive pocket watches, with which they can clock the speed of their horses to a quarter of a second... After an interrupted trip of two days and one night, I arrived here about four o'clock in the afternoon. We had a delay in the woods because the locomotive of the train coming the other way had derailed."

Patek traveled via Columbus (January 31) to New Orleans. There the traveler noted on February 10, 1855:

"Since my letter of January 31 I have had two accidents which, thank God, only cost me time. On the "Alabama", some of the 500 cotton bales with which the ship was loaded caught fire, but the fire was extinguished quickly. On the way from Mobile to New Orleans our boat ran aground. After three days a small boat came to help us. Our boat was floated again, and so we could finally steam away."

But he could not predict rosy prospects for his watches:

"Here everybody complains of the poor business conditions, but everything goes on somehow, and everybody hopes that the rivers will soon be more navigable."

His travels took him on via St. Louis to Chicago. Patek was staying there on February 28, 1855, a good quarter of a year after he had left Geneva:

Man's gold watch with central seconds and date indication, current production. Gold model for men, model 96 ("Calatrava"), made in 1932.

"I left St. Louis on the morning of February 23 and hoped to reach Chicago during the night, but a snowstorm that put snowdrifts three to five meters high in our way stranded us on the prairie for four days and three nights. We froze, starved, and helped the few crewmen to clear the track again. Finally, thank God, I got to Chicago, which I shall leave this evening in the direction of Louisville."

On March 10, 1855, Antoine Norbert de Patek again reached New York. On March 21 he boarded the "Pacific" (which, incidentally, disappeared without a trace during an ocean crossing the following year), with the goal of arriving in Liverpool on April 1.

He wrote to his business partners: "With the same post you will receive a letter from the president of the American exposition, in which you will be notified of the dispatch of the certificate and the medal. Please don't tell anyone about it before I get back. Then we'll publish the letter in the newspapers."

The next letter, dated July 10, came from Paris again:

"The most urgent news I have to share with you is that the rest of our show models absolutely have to be at the show on June 26, because the jury is coming to judge them on June 27."

An important message, as it turned out, for Patek Philippe was honored at this Paris World's Fair with a gold medal. In the coming years Antoine Norbert de Patek continued these trips, which were an essential element in the continuing further development of the company, which was now able to sell more than 1300 watches a year.

About his European trip of 1858, which took him to Moscow, the businessman recorded, for example: "sold Mr. Alexander Pushkin, Lieutenant in the Imperial Guard in St. Petersburg, a hunting cased pocket watch no. 13,549 . . ." The lieutenant was the son of the famous Russian poet, Alexander Sergeyevich Pushkin.

The further history of Patek Philippe ran smoothly, but included, in the remaining years of the century, several thoroughly noteworthy events, decisions and anecdotes:

—In 1863 Adrien Philippe published an inclusive and thoroughgoing book on "The Keyless Pocket Watches, that are wound and set without a key". At this time he also took on the position of the journalistic expert on all questions concerning the pocket watch industry for the Geneva daily newspaper, "Journal de Genève".

—In the same year the management decided that, for the coming months, production at the Grand Quai could be carried on only by lamplight. It was not that daylight had disturbed the watchmakers or hindered their work in any way; the reason was much more that just opposite the building the work on the "Pont du Mont Blanc" Rhone bridge was beginning. Since the work of building the bridge was probably more interesting to the workers than watchmaking, the windows of the factory were covered by big black cloths. The view was spoiled and production was assured. That was important, for in the 1860-1865 period the demand had increased

The "Ellipse" model in skeleton form, circa 1982.

again and nearly 2500 watches a year were produced.
— In 1868 Patek experimented with a wristwatch for the first time. An ornamental watch with baguette movement, to be worn on the wrist, left the factory. The dial was hidden behind a large diamond.
— On November 13, 1871, it appeared that a guardian angel was watching over the factory. On that day a fire broke out that, fanned by a strong wind, destroyed almost all the buildings on the Grand Quai. But the raging fire, that burned for several days, spared the headquarters of Patek Philippe, although both neighboring buildings (two hotels) burned down.
— In 1873, after the lease for the building in the heart of Geneva had been extended for another fifteen years, there was a major change in production procedures. Patek Philippe installed a water-power apparatus on the ground floor, behind the sales rooms. Downright gigantic transmissions of energy were "transported" from the courtyard of the house to the upper stories. This resulted in another increase of production figures, to more than 3350 watches a year, as soon as machines could be installed on almost every workbench, including those for small but time-consuming jobs such as polishing.

The watchmakers of the time were astonished, and rumors abounded. A few years later some strange assertions about the building and the machines were let loose in the world by a British journal:

"Patek Philippe & Co. claim to have the biggest pocket watch factory in Geneva, and the only one to this day equipped with steam power. The advantage of the latter is questionable, to be sure, for the other firms are sufficiently served with water power. The manufacturing facilities themselves are of American origin. The great goldsmithing and jewelry firm of Tiffany & Co. of New York built the factory in Geneva that now belongs to Patek Philippe. Tiffanys believed they could do their own production with the use of Swiss workers and pocket the profits themselves instead of the Swiss manufacturers. This experiment lasted four years, until it clearly showed itself to be a failure. Then the biggest pocket watch factory in Geneva was turned over to Patek Philippe & Co., which from then on supplied Tiffany with pocket watches."

When this article appeared it was above all the technical director, Adrien Philippe, who was very angry. For one thing, because he had invented most of the machines and tools himself, and for another, because Patek Philippe had already been making watches to order for Tiffany for decades. An opposing view appeared two months later:

"Concerning the factory of Tiffany & Co. here in Geneva, only one point is correct in connection with us, and that is that we handle their pocket watch business here and supply the firm with pocket watches. As can be learned from the 'Journal de Genève', we have used machines during the entire past thirty years. As for the steam power that you speak of in your article, we have never employed it, although we could have gained an advantage from it because of our considerable production. But the requirements for using a steam engine are not present in our factory. We have neither

"Ellipse" model with diamond bezel; blue and gold dial decorated with four diamonds.

The "Golden Ellipse" model was created in 1969; the watches shown at left and right come from present production.

ever worked in a building owned by Tiffany & Co., nor have we used a single one of their machines, the greatest part of which were shipped back to America."

In 1876 there were changes in the partnership. Vincent Gostkowski, a member of the firm since 1845, withdrew. His successors were three colleagues in the firm, Albert Cingria and Gabriel Marie Rouge of Geneva and the German Edouard Köhn. The last was a son of Carl Köhn, court watchmaker of the Grand Duchy of Saxe-Weimar; he had joined the firm in 1861 at the conclusion of his studies at the Watchmaking School in Geneva, and had advanced steadily because of his extraordinary ability.

One year later, on March 1, 1877, Count Antoine Norbert de Patek died in Geneva. The title of count had been given to him during the lifetime of Pope Pius IX for his contributions as an active Catholic, also outside the Polish emigrant community. In his place the Frenchman Joseph Antoine Bénassy-Philippe, a stepson of Adrien Philippe, came into the management. The son of the late founder, Léon de Patek, remained a silent partner.

Whoever is so successful on the market must now and then deal with forgeries of poor quality that try to profit from the good name of a brand. The first falsified Patek pocket watch dated back to before 1850, and only one small 'c' too many made the difference in the signature: "Czapeck & Patek".

In 1885, at the Antwerp World's Fair, the jury, of which Adrien Philippe was a member, discovered a forgery that resulted in a court case concerning market rights for watches.

In the showcases of the watch dealer Armand Schwob & Frére of La Chaux-de-Fonds and Paris there was a pocket watch with the signature "Pateck & Cie., Genève". Of course Adrien Philippe, as technical director, knew the watches of the house inside and out. The trick with the extra 'c' was spotted quickly, but the perpetrator remained unabashed. Out of the variety of viewpoints there developed legal action which was to be decided before the cantonal court in Neuchâtel on three days

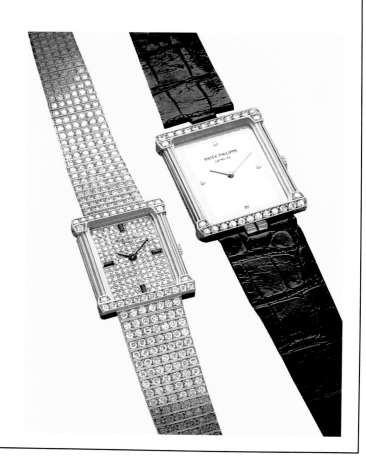

Wristwatches from the current "Les Greques" line. The women's model shown at left center is set with approximately 7.1 karats of diamonds, the men's model at right with 0.43 karats; quartz movements.

of November, 1890. The perpetrators at first defended themselves by saying that they had not built the watch themselves but had bought it in Paris. As evidence to the contrary, Patek presented another gold pocket watch in which was engraved not only "Pateck, Genève" but also "A. S. & F.", for Armand Schwob & Frère. When a 65-year-old watchmaker finally testified that he, as an apprentice for this manufacturer, had had to engrave "Pateck" in the lid of the watch, the case, that concerned perhaps the most extensive incident of watch falsification in the Nineteenth Century, was settled. The cantonal court passed a sentence that was upheld in every point by the Swiss Federal Court three months later. According to it, the manufacturer was not allowed to use the name "Patek" or "Pateck" any more. The number of falsified pocket watches was estimated by the judge and the profit from them calculated at 15,000 francs. Exactly this sum had to be paid to Patek Philippe, and the court costs were likewise charged to the falsifiers. The judgment was to be published in five newspapers at the falsifiers' expense.

In 1888 the company was so strong financially that there were no negotiations for a new rental lease of the house on Grand Quai. Patek Philippe bought the house, had it torn down and a new building built two years later, which is still the firm's headquarters. To reduce the lost time and limit the production to a minimum, more construction was done in the winter of 1890-91, which was not exactly simple. This winter, in fact, was a hard one even for life in Geneva. The contractors had buckets of coal set up on the building site as sources of heat, and the mortar was mixed with alcohol added. Even so, the builders had to stop work for several weeks.

Since the design and arrangement of the building were highly modern and progressive for their time, and offered a view into the production under the best working conditions and the watch trade at the turn of the century, let us take a brief tour of the firm's headquarters as it existed at that time:

Water turbines had been built in the cellar to drive electric generators. With a capacity of thirty horsepower, they supplied the entire house with power for light and for the electric motors, which in turn transmitted power along the ceilings of the production rooms to run the individual machines. A very modern central heating system was also located in the cellar, producing enough heat for the whole 1200-square-meter house. The temperature in the individual rooms could be regulated separately.

On the ground floor were the opulent sales rooms. The largest of them was also the most elegant, its walls being decorated with black and gilded wood and papered with old "Cordoue" leather. A large oval showcase on one wall proudly displayed the firm's medals and awards, and great candelabras hung from the stucco-decorated ceiling. Everywhere in the room there stood heavy bronze figures in the style of the times, surrounded by flower arrangements. The costliest watches were displayed for clients at the sides. Whoever went one story farther up the wooden stairs entered the main offices and a reception room decorated similarly to the sales rooms. Next to them were the business and correspondence offices, the bookkeeping room and the accounting office. This "comptoir" or "reckoning room" was a sort of large-area office by modern standards, in which a partner in the business was always on hand during the working day to oversee the watchmakers and be available to answer questions. Several watchmakers, who in turn worked under the direction of an experienced inspector, were on hand in this room for the firm's quality control. It was their job to watch over every step of production for every single order. Whenever a single piece of work was finished, it was brought to the "comptoir". The next step of finishing was carried out first when a watchmaker, after a thorough examination, was of the opinion that everything about it was right. If there were problems, the piece had to go one story higher, until after additional work it finally passed the quality control.

On the third and fourth floors, each of them a single large room, were the electrically driven machines. The watchmakers sat in long rows at their workbenches, on high wooden stools without backs. Men for the most part, but also several women, dressed in comfortable white or black smocks, worked on generally tiny parts under the loupe, with light falling on their work from high windows. Additional light was reflected by sheet-metal shields from electric lamps hanging on the ceiling. In rows behind the workbenches stood the machines that produced and prepared the parts, linked by a bewildering array of belts and pulleys that connected them with the driveshafts running all along the ceiling. Raw materials in the form of metal staffs or bands were stored nearby. And behind them were large chests of drawers in which additional components such as hands and dials were carefully kept. The fifth and sixth stories looked similar, except that here there were no electrically driven machines as in the so-called mechanical department. Here the parts for the escapements, which were responsible for the precision of the watches, were made perfectly and precisely by hand.

In 1891, when production began in the new building, there were changes in the ownership of the firm. Edouard Köhn withdrew and bought the firm of the Swedish watchmaker Henri-Robert Ekegrén a few houses away. Albert Cingria went with him. The successor to the two of them came out of the firm's own ranks: Antoine Conty, already director of manufacturing at Patek Philippe for several years. Adrien Philippe, last surviving found-

Man's wristwatch with extra-thin automatic movement.

ing partner of the firm, and decorated in the interim by the French with the cross of the Legion of Honor for his achievements, made way for his youngest son, Joseph Emile Philippe.

An essential economic change of the firm's status was made by the proprietors in 1901. On February 1, with a capital of 1.6 million Swiss francs, a stock company was founded, the "Ancienne Manufacture d'Horlogerie Patek Philippe & Co. S.A.". The leadership consisted of seven members, five of them being the owners, namely Joseph Antoine Bénassy-Philippe as chairman of the board, Jean Perrier, François Antoine Conty, Joseph Emile Philippe, as well as the director of the New York office and thus the man in charge of American trade, Alfred G. Stein. After the withdrawal of the silent partner Léon de Patek, there was no longer a member of the Patek family in the company.

As a result of the positive development, the new building had to be made one story higher in 1907-08. By the end of September 1908 the construction work was finished and an electric clock was set in the gable that gave the house its present appearance. Patek Philippe always put much emphasis on the fact that this clock on their facade was controlled by the city clock control of Geneva and that they themselves are therefore not responsible for its accuracy. Just a few years ago it was tied into the circuits of their own quartz master clock, which also controls all the other clocks in the house. The same master clock model is also responsible for keeping exact time in the Vatican.

The completion of the construction in 1908 was celebrated with a big banquet, to which all the employees were invited. A menu was served that included dishes named after characteristic watchmaking activities.

The next years at Patek Philippe were under the influence and gradual spread of a new type of watch, at first very controversial and often described as a silly fad of fashion, the wristwatch. Whether the watches were externally matter-of-fact models for men and women to wear, ornamental watches or complicated timepieces for the wrist, Patek Philippe had a decisive influence on all areas of the wristwatch's history. As early as 1915 a woman's wristwatch with five-minute repeat turned up. In the same year the first wristwatches of the "Chronometro Gondolo" type were created. The movements of these watches were prepared with special features especially for a Brazilian watch dealer and finish the

Gold pocket watch with minute repeat, perpetual calendar, moonphase indication and split second chronograph with 30-minute register. This complex model is still made today.

history of the pocket watches of the same name.

The tendencies of the times, as well as the consequent developments, led to the fact that by the mid-Twenties Patek Philippe was delivering more wristwatches than pocket watches. On the other hand, the effects of World War I, the beginning world economic crisis, plus the firm's internal problems, resulted in considerable business losses, as will be seen in greater detail later.

Nevertheless, in 1925 the first wristwatch in history with a perpetual calendar could be offered. To be sure, the Patek Philippe watchmakers based it on an available woman's pocket watch of 1898. But its springing day, date and month indications were so laborious in technical terms that later models were produced without them. Series production of wristwatches with perpetual calendars began only in 1941. The model 1518 with chronograph was used, which regularly commands high prices at auctions today. Until then, individual pieces with perpetual or "simple" calendars were made, usually to order. In 1925 the production of men's wristwatches with minute repeat was also begun. Some forty such wristwatches, either individually or in small series, were made until 1962, when the last examples were sold.

Until the Thirties the various types of wristwatch movements were available at three levels of price and quality. The differences mainly involved the construction of the escapements and the resulting running results; even the third-quality movements gave more than average performance.

In 1927 Patek Philippe sold the first wrist-

Man's gold wristwatch with "eternal" calendar and moon phase indication; hand wound movement with central second; in production since 1952.

watch with chronograph for 2135 Swiss francs. The crown was also the push-button that activated start-stop-return actions. In the same year the company proved again that it was quite possible to make watches to meet special requests and thus to build movements that had never existed until then. An example is the order of the American automobile executive James W. Packard, who wanted a pocket watch with alarm. That would have been no problem for the watchmakers if the millionaire had not expressed one particular wish. Instead of the unattractive buzzing or ringing, a specific melody should sound, his mother's favorite song, "La Berceuse" (the lullaby) from Benjamin Godard's opera "Jocelyne". On March 8, 1927, the American was able to take possession of his watch for 8300 Swiss francs. It was uncommonly large, with a 29-ligne movement, but Patek had fulfilled the request, and they had also equipped the movement with minute repeat.

The world economic crash of 1929 plunged the once-flourishing business into the greatest financial problems despite the new production of wristwatches and many other efforts. Another reason for the firm's fall was that in the past decades the developments on hand had been put to use, but since the death of Adrien Philippe, the technical "driving force" of the establishment, no remarkable new ideas had come forth, no significant inventions (other than two not overly important patents) had been made. The firm had depended on its good reputation and its loyal customers and, as is well known today, lived for the present. This could go on for a time, but it was not sufficient to overcome a worldwide economic crisis. In 1932 there was practically no money there because, as also happened to other firms, there were scarcely any more customers to be found for high-quality and therefore very expensive watches. For that reason the owners decided to sell the majority of the shares as a way of liquidating the company. The first offer to buy came from Jacques David LeCoultre in Le Sentier. Since 1905 he had supplied most of the raw movements used in the factory. But after much consideration, the men at the Grand Quai in Geneva declined.

Perhaps the second offer already lay on the desk at that time. It came from another supplier who had already worked for the firm for decades: the "Fabrique de Cadrans Stern Frères", and the Stern Brothers, Charles and

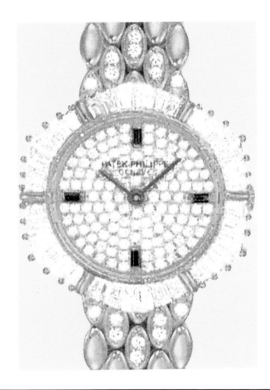

Women's ornamental watches with quartz movements; Seventies. The octagonal model shown at left is decorated with 3.70 karats of diamonds and 0.12 karats of sapphires, the one at right with 1.20 karats of diamonds and 0.97 karats of rubies.

Jean, surely also had an interest of their own in not losing one of their main customers (Patek Philippe used only this firm's dials). On June 14, 1932, the Stern Brothers took over the majority of Patek Philippe & Co. shares, with the result that the grandson of co-founder Jean Adrien Philippe, at that time director of the sales department, had to leave the company and the partnership. Thus there was no longer a direct descendant of the sounders in the company, of which Jean Pfister became the new chairman of the board and technical director. He had formerly directed the Geneva branch of the "Tavannes Watch Co.", that had been closed in this same year. A few months after having been called to this position, Jean Pfister made a decision that had a decisive significance for the further development of the company. He decided that Patek Philippe should once again produce its own raw movements, and that the wares still on hand had to meet the new requirements. The road to a secure future was opened, and the firm revived at increasing speed. In the following year the Stern family involved itself more closely at Patek Philippe. Henri Stern, the son of Charles Stern, founded the Henri Stern Watch Agency in New York with his family's support and took over the American distribution of Patek Philippe watches.

With the renewed production of the firm's own raw movements, an absolute "bestseller" was created, the Model 96, also known as "Calatrava". The cross in this name has stood almost since the beginning as a symbol of Patek Philippe. The wristwatch developed in 1932 became a classic masterpiece in fulfilled, timeless beauty without superfluous decoration, so perfect that even today it is still produced and sold, unchanged outside, only updated time and again in its technical inner being. It was a successful beginning for the new leadership of the business.

In 1936 the decisions of the new management bore their first fruits. In this year Patek Philippe, after the red ink of the past, could show a modest profit again for the first time: 6000 francs. The new chief's next move came at the end of the Thirties. Until then Patek Philippe mounted a great variety of the most varying calibers in its watches. That came to an end. The factory limited itself to a few proved calibers of the highest quality. On the one hand, this substantially simplified the production of watches, and on the other, running precision was raised to its peak (in the latter half of the Forties the first ultra-precise wristwatches with tourbillon were also made), with the successful result that from 1944 to 1966 nearly 500 wrist and pocket watches could compete successfully in the precision contests of the Geneva Observatory.

With economic recovery, technical development also came to the fore again. Patek Philippe improved basic wristwatch techniques and once again became the "best customers" of the patent attorneys in Bern. In the thirty years from 1949 to 1979, forty patents were entered for this company.

The most important will

Man's ultra thin wristwatch with hand-wound skeleton movement, enameled dial ring; end of the Seventies.

Man's gold wristwatch with automatic winding, perpetual calendar, leap-year and moon-phase indication; made from 1981 to 1985.

Man's thin gold wristwatch with automatic winding (movement with microrotor), perpetual calendar, leap-year, moon-phase and 24-hour indication; produced since 1985.

be noted here:
- The Gyromax balance of 1949 made more precise regulation possible.
  Most important was the fact that this balance could also be regulated after being built in.
- A new way of attaching the hairspring to the balance block, in 1958, when Henri Stern returned from America to take over the leadership of the firm as Jean Pfister's successor.
- In 1959 came the "setting arrangement for pocket and wristwatches that allows any desired number of whole hours to be set forward or back without moving the minute hand". Thus a quick and easy changing of local times was made possible for world travelers, since the time zones established in 1870 have only hour differences as their smallest time units.
- In the same year there was also a patent for a "wristwatch movement with straight-line time indication, analog and digital, by two horizontal turning cylinders". The time was read through two horizontal apertures in the case, but the prototype that led to the patent remained a single piece. Patek Philippe never went farther with this type of construction.

At the beginning of the Sixties another threat to the Swiss watch industry and to Patek Philippe arose. With unbelievably low-priced quartz watches and an aggressive sales strategy, the Japanese made a complete shambles of the wristwatch market in a matter of a few months. After long meetings of the directors, in which there was probably much talk of tradition and of thinking toward the future as well, the following decisions were agreed on:

- Patek Philippe would continue to build high-quality mechanical watches in any case.
- Patek Philippe would never build quartz-controlled watches with digital indication via light diodes or fluid crystals.
- Patek Philippe would, however, build quartz watches with analog indication as an alternative to mechanical watches, but they would have to live up to the quality standards of the house.

Thus they could turn back to the experiences of their electronic department, which had been founded in 1948, unnoticed by the public, and which had achieved ground-breaking success as early as 1956. It built the first independent quartz "pendule" and the first fully electronic quartz watch without moving parts. With that, Patek Philippe was already leading in electronic technology at that time, which is shown not least by the high numbers of quartz-controlled master clocks that have been set up in many countries of the world to keep official time. In 1970 Patek Philippe, by using the Beta 21 caliber developed by the Centre Electronique Horloger (CEH), could offer its first quartz watch. Patek Philippe had already linked itself with this research institute several years before by becoming a stockholder. Soon afterward the firm's technicians had developed their own calibers, so that the factory

On the following three pages the seven handworking arts of Patek Philippe are illustrated:

A watchmaker finishing a hand-wound movement; above is an automatic caliber with gold rotor, that is, however, no longer manufactured.

A designer planning a new "Golden Ellipse" model.
A goldsmith, case-smith, polisher.

even owned the thinnest quartz movement in the world (2.5 mm).

Although the production of quartz watches was continued after the formative decisions of the directors, the development of mechanical wristwatches continued independently of it. In 1977 the thinnest automatic movement, with a thickness of 2.4 mm, came onto the market; it is still in production today. In 1985 the company combined this automatic movement with a fully newly-developed perpetual calendar module. The total thickness of the movement is 3.75 mm.

The company has almost four hundred employees today. Production is divided among three factories: in the Vallée de Joux, in the six-story building dedicated on the firm's 125th anniversary in 1964, with its large windows overlooking the Rhône shore in Geneva, and in the old headquarters, where not only sales and administration continue but also the manufacture of quartz watches. An elite corps of selected watchmakers forms the "Atelier des pièces spéciales", in which the tradition of highly complex watches is continued. A "hobby" that keeps the firm at the head of its field, since it is inclined to build its watches in small series rather than bowing down to mass production. Every morning Philippe Stern (born 1938, introduced into the firm step by step by his father since 1977) opens a discussion of division chiefs and directors in the conference center on the second floor of the main house. Every day the most important letters and orders are examined and problems are discussed. Philippe Stern is now the general manager of the firm, while his father serves as the president. Both of the firm's owners have the ambition of continuing to run Patek Philippe as a family business.

Seven handworking arts guarantee that the production of watches from their own raw movements and from ebauches, mainly from LeCoultre, will be maintained in the tradition of the highest requirements of quality:

— *The Watchmakers:* Whoever has reached the level of master in this field comes to Patek Philippe only after taking a second "course of training" at the factory. The "hand" workers, who realize that they are the protectors of tradition and perfection, have carried on the development from Adrien Philippe's first watch with crown winding to electronic watches and brought forth the first fully electronic watch as well as the first miniature quartz chronometer.

— *The Designers:* They are artists and artisans all in one and must live with the fact that in the end only one model may be chosen from a hundred designs and prototypes. Their slogan: A watch is certainly a useful object in the broadest sense, but a Patek Philippe must fulfill the further task of setting off the wearer as someone special.

— *The Goldsmiths:* They are the house smiths, precious stone setters, polishers and above all masters of their trade. Even in the Twentieth Century they still work with the methods of over

A chainsmith at his work on a gold band for a wristwatch.
An engraver chasing a wristwatch case.

a hundred years ago, and with correspondingly traditional tools.

- *The Chainsmiths* With only a few tools and mostly by hand, without the use of machines, they cut, plait and weave delicate gold threads into perfect watchbands and fabrics of gold. And to be able to work as a chainsmith at Patek Philippe, the applicant must have finished a further four-year "perfection" training after his regular four-year course.
- *The Engravers:* They engrave, chase, and punch with unbelievable calm and perfection. The difficulty of their work consists of working the precious metal, gold, which the specialists class as one of the "living metals", as it never behaves quite the same under the artisan's tool.
- *The Enamelers:* There are only a few people left on this earth who have mastered the art of transferring complete paintings onto a few square centimeters of surface. The tiny details are often painted on with a single hair. They work so perfectly that Patek Philippe is the only firm in the world capable of delivering an enameled watch with any motif that the customer wants.
- *The Jewelers:* As with all the specialists of the handworking arts, the jewelers also must follow a long path before they can work at the decoration of Patek Philippe watches. As precious stone specialists, they choose every jewel individually according to its color, purity, weight and size. They set the stones and polish the gold. Thus every watch, along with its absolute precision, takes on the pure beauty of a piece of jewelry.

Preservation of tradition and firm production principles have ensured that every single part of every watch will be worked by hand; even the finest screw is finely polished it can be put into a movement. The assembly of a single watch takes at least nine months. And there is hardly an upper limit to production time. Even five years' work in the production of a complex wristwatch is quite possible today.

These facts and the limitation of daily production to at most fifty watches help to explain the high sales and collecting value represented by Patek Philippe watches today. Every watch is really an individual piece, even when it comes from a series, because one watch can never be identical to another to the slightest detail.

In the realm of wristwatches, various lines are offered today, among which the aforementioned "Calatrava" is the oldest. Created in 1932, this symbol of the most perfect beauty has not been changed since. Patek Philippe explains this fact in a masterpiece of understatement: One would think just as little of changing the proportions of the Venus de Milo. The watertight or dampness-resistant model is made in yellow or white gold, with hand or automatic winding or quartz movement. The proportion of mechanical to electronic watches is presently about 60:40 at Patek Philippe, wherein the quartz watches are mainly women's models. The timeless line is represented by the "Golden Ellipse", born in 1969. Its proportions are based on the golden section by which the Greek temples and the Cathedral of Notre Dame were built. To emphasize its clear beauty in mathematical harmony, the factory has created a blue-gold

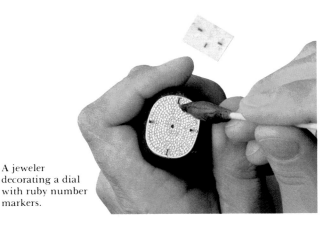

A jeweler decorating a dial with ruby number markers.

An enameler finishing a miniature in enamel painting.

dial.

Since then the "Golden Ellipse" has also been made with white, gold, and diamond-set dials, although the blue-gold creation is by far the most popular. Hand-wound, automatic and quartz movements are likewise available. The most unusual variation of this model is surely the skeleton wristwatch with a mechanical movement. At this point, and with this watch as an example, it is time to refute the legend that a genuine Patek Philippe can always be recognized by the Calatrava cross on its golden crown. The "Golden Ellipse", for example, was at first made with a crown decorated in that way. Later, though, it was decided that this decoration was inclined to disturb the complete design. So the watch was equipped with a new rounded crown without the Calatrava cross.

A porthole provided the inspiration for the creation of the "present-day line", the "Nautilus", whose case is made of a block of stainless steel or gold and thus guarantees absolute watertightness to 120 meters for men's and 60 meters for women's watches. Every "Nautilus", with the exception of a few models set with fewer diamonds, indicates the date and is protected by a scratchproof sapphire crystal, almost as hard as diamond. The mechanical movements without exception have automatic winding.

In addition there are various so-called "fantasy lines", which offer the most varied possibilities in thoroughly exclusive design. Finally there are also complicated wristwatches in the Patek Philippe program, which can be described briefly:

— *The Model 3940* with automatic winding, perpetual calendar, moon-phase, leap-year and additional 24-hour indication. It has already been mentioned that the movement and calendar mechanism together have a thickness of only 3.75 mm. With the presentation of this fully new model in 1985, Patek Philippe continued its long tradition of wristwatches of this type.

— *The Model 1970*, which followed in 1986, and whose production run, not yet resumed, was sold out in the shortest time. This is a new version of a classic wristwatch with chronograph and perpetual calendar. The model was also enhanced with leap-year and 24-hour indication.

Gold ornamental watches, circa 1980.

Gold wristwatches from the "Nautilus" line; at left a woman's model, watertight to 60 meters, quartz movement with date indication; at right a man's watch, watertight to 120 meters, automatic movement with sweep second and date indication; present production.

— The high point of the watchmaker's art in the world of wristwatches is certainly represented by a unique creation of 1986. Along with chronograph and perpetual calendar, it also has minute repeat. For the time being, Patek Philippe is not thinking of selling this watch.
— Time-consuming and complicated to produce are also the various *Skeleton Models*, in which the movement can be seen in all its transparency.

In 1989 Patek Philippe will be 150 years old, an anniversary that in all probability will offer a surprise to the admirers of this brand. It can be expected that Patek Philippe will fill out its collection with a more complicated wristwatch.

In conclusion, though, let us take up the question of who the clients of Patek Philippe are. In any case they are people who know something of precision watchmaking and have a feeling for quality. A part of them, the smaller portion, "earns" a Patek Philippe, which means nothing else than that the watch must be saved for, and that its purchase fulfills a long-cherished wish. The other, more numerous portion generally has sufficient financial means at its disposal to be able to buy what its heart desires without great sacrifices. Prominent people in all realms of public life often belong to the latter group. But Patek Philippe says not a word about them, for discretion is part of the business. Often it is not known in Geneva when a Patek Philippe is placed on a prominent wrist elsewhere in the world, for the retailers are also discreet. In the green archive books that are surrounded by secrecy, only such personalities are listed as final purchasers as bought their watches at the firm's own shop. Among them are Walt Disney, Albert Einstein, Ella Fitzgerald, Charles Lindbergh, Artur Rubinstein, Peter Tchaikowsky and the Duke of Windsor. Queen Victoria, who obtained her watch directly from the firm's founder at the London World's Fair, has already been mentioned. The notation "stolen" is often to be found in the archive books, a sad indication of the desirability of this house's watches. With the entry goes the hope of being able to return the watch to its owner if it should ever find its way back to Geneva and to the service department there. On the other hand, it must be stressed that Patek Philippe builds watches that constantly increase in value and withstand the pressure of time unharmed, by virtue of their precision, watches to be inherited, watches as a capital investment. How much the value of Patek Philippe watches can increase is shown more than impressively by the results of auctions in the past year. An increase of 100% and more within a few years is not uncommon, and for watches with minute repetition, nearly 1000% has been attained. This puts the apparently high original price into perspective.

Another factor is that the principles of manufacturing have not changed in the almost 150-year history of the firm's existence. Patek Philippe has remained true to its founder's guidelines. That is the guarantee that today's watches with the sign of the Calatrave cross will remain the masterpieces of the best handworking art.

Gold "Nautilus" for men, decorated with diamonds and emeralds, movement with automatic winding.

PIAGET

It was like an Oriental fairy tale. On the main floor of the Piaget watch factory in Geneva there sat a sheik, wrapped in white Arab garments, swept the table with a judging glance, and worked his way very cautiously through the shimmering gold and glittering gems of the wristwatch collection that lay before him on soft brown velvet in elegant hardwood boxes. He put a few of the most valuable pieces aside. But then the bearded, sun-bronzed man seized one of the sparkling items and made it known that, in addition to the other watches that he had already chosen, he wanted this piece most of all.

No problem!

But the wishes of the Oriental ruler, who was accustomed to having his innermost thoughts responded to at once, were not yet fully satisfied. This one watch was beautiful, but he needed thirty-one more examples in women's size for his harem. No problem!

But the sheik's wishes still seemed to have no end. All thirty-two watches had to show differences, for he could not give each of his ladies the same watch. Now there was a problem!

But Piaget would not be Piaget if they could not handle such difficulties. At any rate, the sheik was assured that his order would be filled to his satisfaction. A designer of ornamental watches was commissioned to travel to the customer's homeland at once and come back to Switzerland with the right idea.

The first surprise awaited the man from Geneva right after his arrival. How was he to tell thirty-two women apart when he could see the female population of the sheikdom only heavily veiled and wrapped in garments that came up to their faces? Did they have different tastes or hobbies, he asked. But the answer got him no further: the ruler's ladies were allowed to have only one preference: for their lord and master and his inclinations and wishes. Could he see their rooms? The artist asked the sheik if he might see the harem, and his request inspired first shock and then hasty activity. The rooms in the palace in which the ladies lived were cleaned out. The Swiss was the first European ever to take a tour of the harem, constantly eyed with mistrust by the sheik's bodyguards.

And he found the "philosopher's stone". All the rooms were differently decorated, and every room had a different costly, fine-patterned wallpaper! When he modestly asked whether he could take a photograph of every room, this "demand" was categorically refused by the ruler. Within a few hours the designer had to copy the designs on paper with a pencil.

When he arrived in Geneva, thirty-two different dials were designed and produced. The order was filled! The sheik could give each of his ladies the same gold, diamond-set watch, individually matched to the wallpaper pattern of her room. One can only wonder what will become of the watches if the sheik ever wants to relocate his harem.

We began this chapter with this anecdote deliberately, because Piaget represents the fairytale quality of Swiss luxury wristwatches, artistically unique, and perfectly and precisely

Yves G. Piaget,
Vice-President

produced. "Nomen est omen," one is tempted to say when one knows that the company had its origins in La Côte-aux-Fées, which means "the hill of the fairies".

Georges Piaget was the first member of the family who, like other farmers in the village, began to work during the long winter nights patiently and precisely preparing parts for watches and putting them together. At an elevation of a thousand meters, where the fields and pastures bear only a meager harvest, such work at home became the main occupation and finally led to agriculture being given up. In the old farmhouse Georges finally assembled the first watches of his own in 1874, and they were sold mainly in the surrounding villages. He engraved "Piaget & Co." on their dials. He himself, Georges Piaget, was "Piaget", and his family was "& Co." This engraving on the dials marks the birth of the firm, a few years before the name was entered in the official register of brands.

Production went on in modest quarters in the next few decades. Nothing remarkable is recorded in the annals of the company. The founder's sons brought the small company through the two World Wars by the skin of their teeth. Things went right with their watch production only after World War II had ended. Then the founder's grandsons, Gérald and Valentin Piaget, took over the leadership of the business. And with them the more or less planless manufacture of watches came to an end. The two young businessmen were very ambitious and turned the more than sixty-year history of the firm completely around. They planned and built the first real watch collection of the House of Piaget. Their type of business leadership was obviously right, for the watches from La Côte-aux-Fées suddenly became very respected and desirable in Switzerland.

In order to continue the growth of the family business, the Piagets extended their feelers into the international market. Thus they laid the foundation of a development that was to carry Piaget very quickly to a high place in Swiss watch manufacturing. The two brothers' success, that motivated their colleagues to their highest achievements could be summed up as follows:

— In 1956 Piaget revolutionized women's watch fashions with the world-famous ultra thin, nine-ligne "9P" movement. Thus it became possible for the older generation to wear an elegant, thin wristwatch for the first time, but one that was big enough so that they could read the time without having to put glasses on.

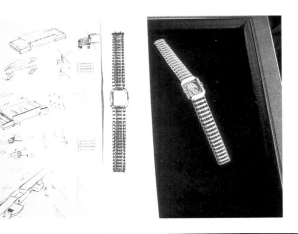

A new wristwatch model comes into being.
Man's gold wristwatch with link band.

- In 1959 Piaget opened its own jewelry shop in Geneva that was known within the firm as the "show-window for the world".
- In 1960 the firm surprised its older, more traditional Swiss competitors with its own development, the thinnest automatic watch movement in the world. The Piaget caliber "12P" was only 2.3 millimeters thick.
- In 1961 the "March to Germany" took place. On March 8 the German Piaget GmbH was founded in Offenbach. It was the firm's own first foreign branch, and began as a two-man undertaking that was meant to make the wearing of Swiss luxury wristwatches tasteful to those West Germans spoiled by the Economic Miracle. The experiment was successful, for at the twenty-fifth anniversary of the branch firm there were already seventeen employees (including six watchmakers and a goldsmith) serving forty retailers in the Federal Republic.

With fewer than six hundred watches sold (the official statistics say 550 to 600), the branch had a gross income of around five million marks in 1985. These figures become more meaningful when one knows that the least expensive Piaget watch costs 5000 marks, with practically no upper limit. In the same year, the main firm in Switzerland had a gross income of around 100 million Swiss francs.

To be able to understand the further development of the firm in the Sixties, after the founding of the first foreign branch, one must know what was undertaken by the Piaget family in Switzerland.

Almost simultaneously, Piaget bought up several gold case and band factories in Geneva. When the family members then had the new work-force statistics reckoned up, there were a few amazed faces. In the realm of high-quality watch manufacturing, the jewelers and goldsmiths suddenly outnumbered the watchmakers, which very soon made it clear that the Piaget collection was centered around jewelry. The firm's purchases and business decisions under the leadership of Piaget led to an autonomy very rare in the watch business. Now the production of the interior (the technical part) and the preparation of the exterior (the decorative part) were both in one hand and thus under absolute control, and could be influenced at any time. This "fantastic self-sufficiency", as it was regarded in Geneva, allowed the gold and watch factories to carry on independent research. And this in turn formed the foundation for the highest achievements, with which Piaget "astounded" its competitors again and again.

- In 1964 the firm almost completely took over the firm of Baume & Mercier, and the Geneva jewelers created a completely new line with jeweled dials. At one stroke, this idea influenced the world's jewelry industry. The "faces" of the wristwatches, set with lapis lazuli, coral, malachite and tiger-eye,

Man's skeleton wristwatch with automatic winding, perpetual calendar and moon-phase indication.

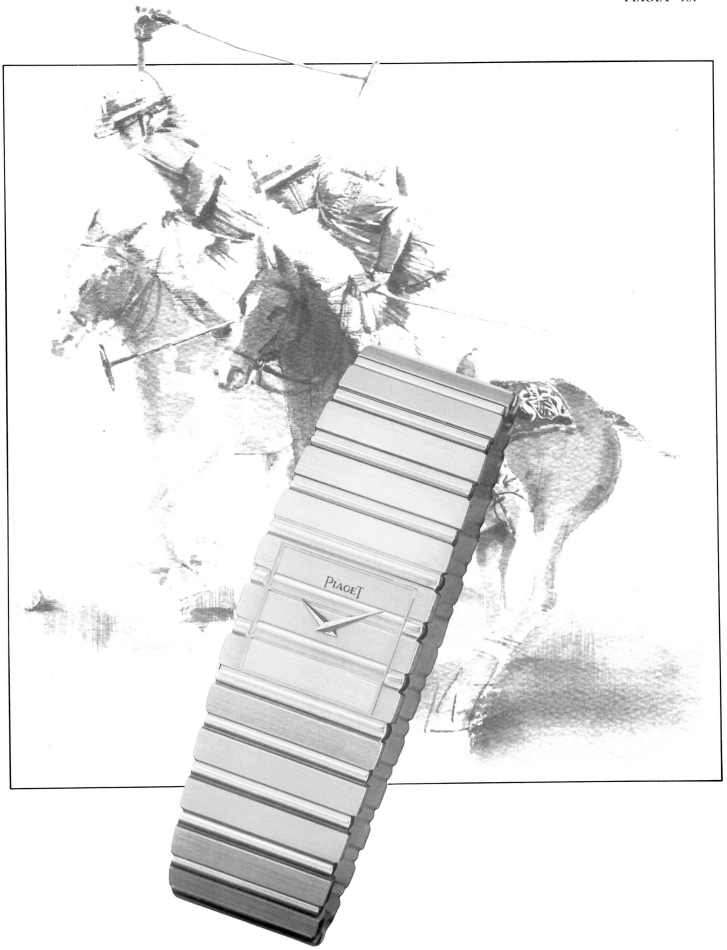

Gold "Polo" model.

thus did not remain exclusive for long.

- In 1969 Piaget built its first quartz watches. The firm was then one of the leaders in the new technology, because it was one of the seven stockholders in the C. E. H. (Centre Electronique Horloger) in Neuenburg. Most Swiss technical development in the realm of time measurement was done there, and Piaget was involved from the start.
- In 1976 the unhindered access to the new developments in quartz technology at this "idea factory" paid off for its owners. Piaget developed the basic discoveries of the eighty scholars at the C. E. H. further in its own research, resulting in the thinnest quartz movement in the world, the "7P Caliber", the only movement in the world with a "memory of the exact time". The watch could be shut off to save battery power, and when turned back on would automatically set itself to the right time. Immediate time-zone changing was also possible by using a special switching device.
- In 1981 the "8P Caliber" appeared, an even smaller quartz movement, this time the only one in the world equipped with electronic time setting and time-zone changing.
- In 1982 the watchmakers at the factory proved with the "Caliber 20P" that they could win a place at the top of the mechanical realm. The movement, 1.2 mm thick, was the thinnest mechanical movement with hand winding. Its reference numbers, unlike those of other movements, began with a 2.
- Naturally the automatic watch also had to be "shrunk". Thus in 1983 the world's thinnest automatic movement, the "Caliber 25P" (reference numbers with 5), appeared. It was just 2 mm thick and outshone the firm's own "Caliber 12P" that had been unbeaten since 1960.
- In 1986 came a new outstanding achievement in mechanical watches: the movement with perpetual date indication through a window in the dial.

But it was not only these masterful technical achievements that established the firm's excellent reputation. It was equally the artistic ideas of the jewelers and designers, who created individual watches in small series, as has already been noted in reference to the development of dials set with jewels. Today Piaget is justified in claiming the title of "Watchmaker to the World", for the firm has given the whole industry impulses that have pointed the way with its impressive creations. During the past decades, no other firm has so imperturbably turned wristwatches into high-quality jewelry and thus secured its own present and future. These "haute couture" wristwatches

Man's wristwatch with quartz movement; case and band are decorated with diamonds and onyx.

Man's octagonal wristwatch, its bezel set with diamonds.

Model with gold band; onyx dial decorated with diamonds.

have naturally led other manufacturers to adopt ideas from Geneva. For that reason, Piaget watches have been copyrighted for years.

But in the case of the "Polo" collection, with its new horizontal-line design, a German manufacturer, who otherwise had a good reputation, paid no attention to this. It took a court's verdict to persuade him that the design of a wristwatch is based on an artistic creation that is protected by law.

It was not just the design of Piaget watches that was copied; complete falsifications also appeared on the market. The "best" years for counterfeiters were 1983 and 1984; the fight against them cost the firm about 1% of its yearly income and at times took up as much as a third of the management's work time. The professionals were located in Hong Kong, Germany, Italy, Switzerland and France. Now the problem has been dealt with. And in Geneva Piaget demonstrates, again with a "Polo", that the difference is in the weight. One need only take such a watch in one's hand. The real Polo has the weight of a small gold bar, but one scarcely feels the counterfeit in one's hand. It is no wonder, for Piaget is the only watch manufacturer in the world today that uses eighteen-karat gold exclusively to "dress" the movements—as they modestly say in Geneva. Only one exception is made when especially expensive gemstones are to be set. Then the jewelers employ another costly metal: platinum.

Two and a half tons are used per year to make cases and bands for timepieces. Five tons per year are purchased, and the unused "leftovers" are not melted down at the factory, but returned to the banks. The dials are also made of gold, with the exception, of course, of those made of stone, and likewise—even if they appear black or brightly colored—the hands, crowns, and clasps for leather bands. In the last case, Piaget recommends that customers make sure that the original clasp is used when the band is changed.

The watch artists of Geneva were and are cautious when buying their precious stones. Whoever buys, as does Piaget, about 7000 karat (in down-to-earth language, about three pounds) of these carbon creations can afford a bit of pedantry, and is in fact obligated to.

"Truing" a balance with the help of poising calipers.

Balance and hairspring are installed in a skeleton movement.

Fine work on the middle part of a case.

This claim to quality naturally shows itself in artistic creation too: the designers have to accept the fact that only about every fiftieth design for a wristwatch will be judged by the head of the family to be worthy of being used as a prototype. Whether it is included in the collection in the end is not yet guaranteed, though. This too is decided by the Piaget family, in conference with their main retailers, who meet once a year in Geneva to choose the new collection.

The fact that for more than twenty years all watches have been fitted with sapphire glasses that are almost as hard as diamonds has long been taken for granted, and is no longer particularly emphasized.

In addition to their highly precise movements, two other components of Piaget watches deserve special attention:

—the gold bands with which the firm got

Various gold men's wristwatches.

away from the usual features of the strictly functional years ago, and
— the dials, the "faces of the watches", which have become veritable cult articles.

If we look first at the gold watchbands, we see that their development began at a time that only the oldest jewelers can remember. At that time the manufacturers attached either their own or purchased bands to watches more or less indiscriminately. The function mattered much more than the design. Piaget inspired a change in this viewpoint. The relationship between the size of the case and the width of the band began to be considered. Cases and clasps were engraved in the same pattern as the band, and these considerations gradually spread to the dials as well. Thus the individual personality of the wristwatch was created, and the watch soon became a piece of jewelry that was designed according to the influence of fashion as well as the wearer's taste. Such gold watchbands were always completely designed and built in Piaget's Geneva workshops. Even the necessary tools were made by the house's own specialists. In the last decade Piaget has slowly adapted all link watchbands to a screw system, so that the bands can be shortened or lengthened by a dealer without problems.

Piaget makes an even greater claim to fame with its dials. The firm's catalog offers an overwhelming assortment, proceeding from the maxim that the "face of the watch" is the part that can best be matched to its owner's personality.

The dials are made either of gold, of semiprecious stones, or of a combination of these materials. The gold dials consist of a 0.3 mm thin gold plate that can be supplied with a white, black or colored finish. The stone dials are cut from semiprecious stones of especially

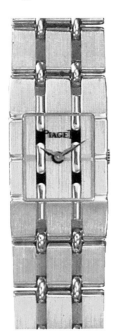

Various gold ornamental watches, some with diamond decor.

chosen quality. Their surface is generally polished, but they can also be given a satin finish or engraved, for example, in the same pattern as the glass rim or the gold band.

The many possible combinations allow apparently "individual pieces" to be ordered. In this way the factory can presently offer more than two thousand models, put together essentially from sixty different cases, sixty watchbands and sixty dials. In principle, Piaget builds any watch that a client wants to buy, but this exclusivity naturally has its price, which may even frighten off an American oil baron. A Texan multimillionaire, after a long struggle, finally decided to buy a watch that until then had been built only ten times, with just the dial changed. This was decorated with his portrait.

The firm will even refrain from engraving the Piaget signature if the client wants it that way. Of course no name at all then appears on the dial. Such Piaget wristwatches are not nameless even so, for "Piaget" is engraved at least in the bottom of the case.

About twenty thousand watches leave the factory every year. About a decade ago, the United States and the Middle East were the most important markets of the house, but since then Europe has pushed the sheiks down to third place. The ratio of men's to women's watches is described in Geneva as "balanced".

Piaget had no trouble making a decision for quartz or mechanical

Left: ornamental wristwatch, set with 724 diamonds adding up to 23 karats.

Man's wristwatch with day, date, month, and moonphase indication, movement with automatic winding, hand polished gold case.

movements, since the factory is a leader in technical development. Every fifth watch still has a mechanical movement, and one gets the feeling that this proportion could grow again. The sales director explains the firm's position as to this "dilemma" by imagining a person about to buy a relatively expensive watch: "The modern person is likely to choose the quartz technology, saying that man did after all give up the coach for the automobile. A conservative customer will give preference to the traditional watchmaker's art, accept a very slightly inferior running regularity, and say that neither he nor his children or heirs ever will want to be dependent on the availability of the right kind of batteries, and that old coaches are worth a lot of money today. From the standpoint of the collector, today's ultra thin mechanical movements are worth a lot more just because they can be produced only in limited quantities."

The registration of watches that are sold is based on tradition now as then, done in handwriting by the watchmakers and not entrusted to a computer. A Piaget is marked with two numbers. The first number is the design identification or reference number, the second the series number. The movement number is written in "petite noire", because it was once recorded in a black leather-bound book and is still known by that name though the book is now green. In a year a watchmaker writes a whole book very neatly, making it possible for the factory to give information as to the date and place of sale upon request from a client. The branch office keeps its own register, in which the supplied retailers are documented.

Along with service to clients, Piaget gives another reason for registration: the counterfeit market. In principle, a Piaget is supposed to have the same price everywhere in the world. For that reason, retailers who vary prices are no longer supplied. The quality of service can be relied on by Piaget owners for decades to come. Any watch can be repaired when it is

Left: gold pocket watch with skeleton movement, complete calendar and moon-phase indication.

Right: man's gold wristwatch with complete simple calendar and moon-phase indication. Hand-wound movement.

188 PIAGET

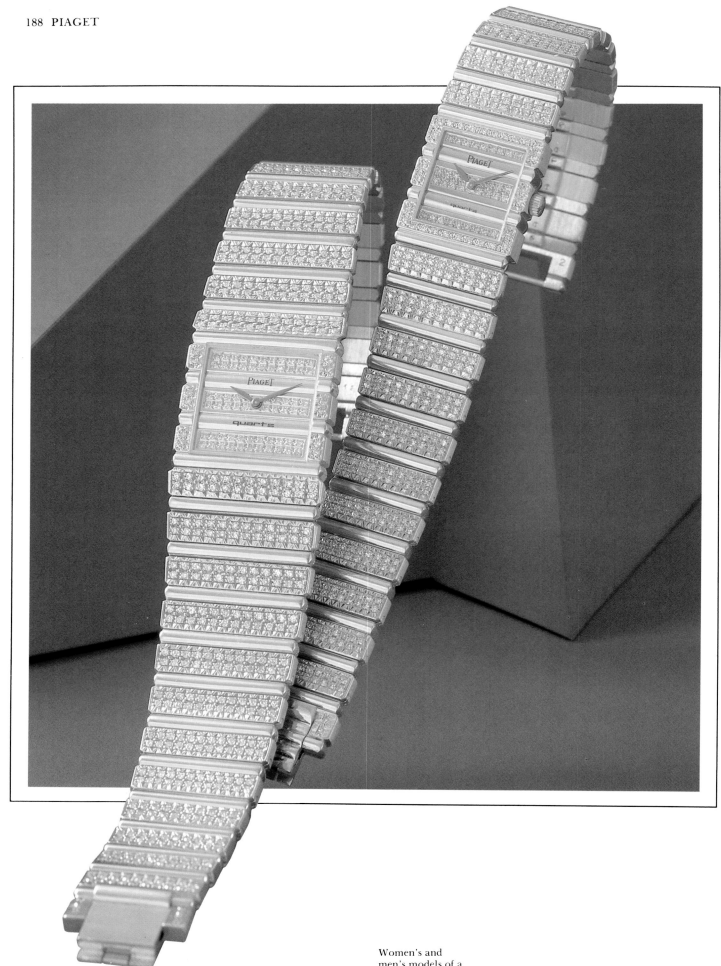

Women's and men's models of a gold wristwatch with quartz movement and diamond decor.

necessary. The vice-chairman recently stated to English dealers the six special features of Piaget watches:
— the exclusive design. They simply look different from others and have a classic atmosphere about them.
— the outstanding beauty and unchanging style. They are a degree more refined.
— the extraordinary quality of their movements, made only for Piaget in the firm's own workshops.
— the extraordinary quality. Every single part is perfectly made of the best available materials.
— the worldwide prestige, since Piaget owners are members of a certain cultural, financial, artistic or political elite and have an inborn taste for fine art and a flair for style.
— the worldwide service network. The best-known watchmaking families, their businesses, many of them more than a century old, and the world's leading jewelers stand behind this product.

With so much self-confidence, the price factor is scarcely of any significance today, thanks to the firm's international success and recognition. Even so, there are seven answers ready:
— the relatively small numbers of each type produced,
— the outstanding quality of every single part,
— the "limited editions" of Piaget movements,
— the quality of the case, cut out of a solid block of 18-karat gold (by cautious, intricate construction methods) to guarantee complete protection to the movement,
— at least three hundred hours of the most careful work by more than three dozen different groups of artisans and experts go into each Piaget watch,
— every Piaget is a work of art—in either its movement, its aesthetic conception, or both, and finally:
— Piaget watches are outstanding luxury products, and genuine luxury has always had its price.

That the "rarities of tomorrow" will maintain their position on the market is guaranteed by the fact that Piaget is, and always has been, a family business. Precisely because the decisions are made in the closest family circles, the firm can react quickly and also unconventionally to the market—another concept that brings success to this firm that, with around five hundred employees, is only medium-sized by Swiss standards.

A parting thought from President Yves Piaget: "You don't read the time from a Piaget—you admire it . . ."

Man's octagonal wristwatch with onyx dial, its center set with diamonds.

ROLEX

Anyone who delves into the Rolex firm's history is search of a "Mr. Rolex" will be disappointed, because Rolex is a word invented by Hans Wilsdorf, the founder of "Montres Rolex S. A." This Hans Wilsdorf came into the world in Kulmbach, Franconia, on March 22, 1881, as the second of three children. He was to revolutionize the Swiss watch industry and have a remarkable career.

When Hans Wilsdorf was twelve years old, his mother and then his father died. Their uncle cared for the three children, closed the business in Kulmbach that the grandfather had founded, and put the children in a boarding-school. During his schooling, Hans Wilsdorf felt drawn particularly to mathematics and languages, and he recognized a longing to travel in foreign lands. After his apprenticeship in an exporting firm, he ventured a first step into the outside world: in 1900 he moved to the Swiss city of La Chaux-de-Fonds.

For eighty francs a month he worked as a correspondent in English and a general-work employee of Cuno Korten on Rue Leopold Robert. In this firm, already quite a significant one, with a yearly income of a million Swiss francs, Hans Wilsdorf first came into contact with the later course of his life: watches. Cuno Korten exported timepieces, only a few of which were made by the firm itself. But this was the best opportunity for the young German to get acquainted with the watch business and all its products. The watches must have fascinated Hans Wilsdorf immediately. The nineteen-year-old showed what he had in him by a courageous move: he ordered one pocket chronometer each from what he considered the three best watchmakers and applied for official running certificates from the Neuenburg Observatory. This inspired his employer to give the young man more responsibility. The course of his life was now plotted. Watches would never let Hans Wilsdorf go. Three years later, when he packed his bags and traveled to his new employer in London, at that time the commercial center of the world, he remained true to his chosen profession. He earned his living working for a watch importer.

Wilsdorf, who had found his particular attraction to watches of high quality in Switzerland, came to two important conclusions in 1905: quality ought to succeed quickly on the English market, and the future belonged to the wristwatch. He was quite alone in the latter view in 1905, when he used his siblings' inheritance to set up his own business (his own 33,000 gold marks had been lost in transit to England), founding the watch-importing firm of "Wilsdorf & Davis" in London. Hans Wilsdorf described those times in his memoirs as follows:

"For fourteen years I bore the financial responsibility and was at the same time the manager of the business, which was a full success from the start, thanks to our basic principle: to deal only in specialties of the watch industry, especially with new articles."
. . .

Hans Wilsdorf
(1881-1960)

"At first it was the luxury leather traveling watch, called the briefcase watch, that I distributed large quantities of in 1905, in a wide variety of types and styles. But my efforts were devoted most of all to the wristwatch; I devoted all the energy of my youth and unquenchable optimism to it." . . . "At that time the style of wearing a watch on the arm was not only not modern, it was even ridiculed because it simply was not seen as compatible with the image of manliness. Then too, the watchmakers of all lands were skeptical, and predicted the complete failure of the wristwatch." . . . "Among others, they introduced the following arguments: 1. The mechanisms of such watches, which had to be small and delicate, would not stand up to the vigor of human activity. 2. Dust and moisture would quickly destroy such a mechanism, even if it was built very solidly. 3. Such a small movement could not possibly be precise or run regularly."

But Hans Wilsdorf was very determined to prove all the skeptics wrong. His first step was making contact with Hermann Aegler in Biel. He had kept this watchmaker's name in mind since his days in La Chaux-de-Fonds, because Aegler could offer small raw movements with lever escapement at affordable prices. This firm had been founded by Jean Aegler in 1878. In its workshops in the "Rebberg" of Biel those eleven-ligne lever movements were made for which Wilsdorf traveled to Switzerland not long afterward, for they had the reputation of running especially precisely and being easy to get replacement parts for on account of their industrial manufacture. At Aeglers, who sold their own watches under the trade names of Final, Precision and Zetex in Germany, France, Greece, Great Britain, Italy, Russia, Scandinavia, Turkey and even overseas, Wilsdorf ordered several hundred thousand francs' worth of wristwatches. This amount was more than five times the value of his company. In London he employed watchmakers who carefully tested the delivered wristwatches in his own studios before they went on sale.

Wilsdorf's visits to Biel were repeated. The knowledgeable purchaser brought new ideas with him, which resulted in hundreds of models. In the entire British Empire, in the Far East, in Australia and New Zealand it became the fashion to wear a watch on the wrist.

His first models were made of silver, with leather bands for men and women. Gold models came somewhat later, equipped with the flexible band invented in 1906. In this way Hans Wilsdorf had made a decisive contribution to the general spread of the wristwatch.

In 1908 the firm ranked among the leading watch merchants of Great Britain, and Hans Wilsdorf wanted to give his watches a name that could appear on their dials. This intention was a frontal

The first wristwatch to receive a "Class A Kew" certificate, and the document that went with it.

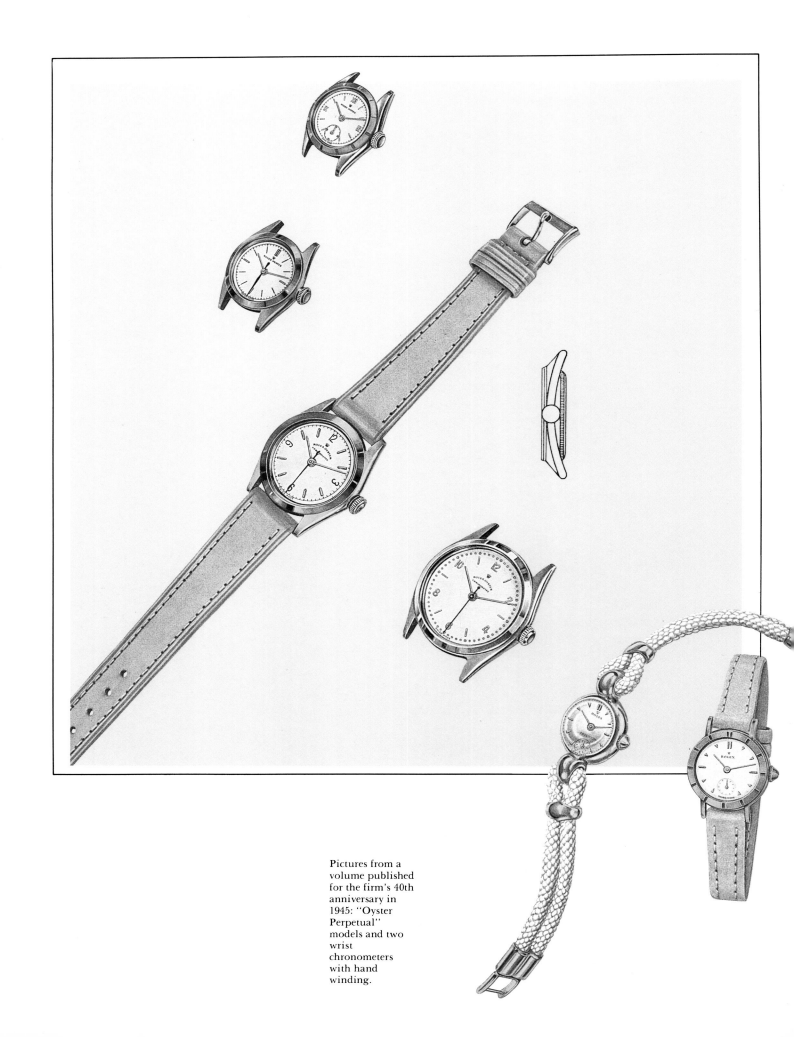

Pictures from a volume published for the firm's 40th anniversary in 1945: "Oyster Perpetual" models and two wrist chronometers with hand winding.

Two square wrist chronometers, and "Oyster Perpetual" chronometer with date indication for men, and an ornamental savonnette watch for women.

"Oyster" hand-wound models for men and women, as well as two gold women's watches.

"Rolex Precision" men's wristwatches and a gold wrist chronometer with hand winding.

attack on British tradition, for since the Eighteenth Century, watch manufacturers had immortalized their own names on the dials. But Wilsdorf wanted to make his watches known by a name that would create a brand.

A first important step was the name itself. Rolex, it is said, was made from "horlogerie exquise". It was impressive, pronounced the same in all European languages, and so short that there was still room for the watch shop's name on the dial. The brand became known gradually. One factor was that its first successes were striven for in terms of running precision. In 1910 came the first official recognition from the Bureau Officiel in Biel.

The greatest success of the time was registered on July 15, 1914, when for the first time a wristwatch of 25 mm diameter (11-ligne movement) was given a Kew-A Chronometer Certificate after 45 days of strict testing at the Kew Observatory in London. This was without a doubt a sensation at the time, as this small watch movement was equal in precision to the marine chronometers. Running precision was tested in five positions (crown up, crown left, crown right, flat on its case bottom and on its dial) and at three temperatures (air temperature of 18 degrees Celsius, in an icebox and a regulated oven). Wilsdorf and his Swiss watchmakers had begun to prove that small movements could also possess the qualities of chronometers. Now he requested of his supplier in Biel that from then on all Rolex calibers should have to pass the chronometer tests. The Kew Observatory attested that this was possible on June 14, 1925, with the first Class A certificate for a woman's watch of 13 by 23 millimeters.

Among the subsequent chronometer successes were a Class A Kew certificate in 1927, with the note, "especially good", for a rectangular Rolex movement (24 x 15 mm). A year later the same caliber received the first first-class certificate given by the Geneva Observatory, where the testing was done under the conditions for pocket chronometers. The list of successes continues to the present day, and it is clear that Rolex has stressed series production of wrist chronometers from the start. In order to achieve this, an incredible degree of precision was needed in the production of the movements, as computers and ultra-sensitive measuring instruments were not yet available then. The efforts paid off. In 1936 a first series of 500 "Prince" caliber movements received the certifice of the watch testing bureau with the notation, "especially good". In 1955 the jubilee of the fifty-thousandth chronometer's production in Biel could be celebrated.

The second step in making the Rolex brand

The title page of the "Daily Mail" of November 27, 1927, with a picture of Mercedes Gleitze and an advertisement for the "Oyster" models.

The "Luminor Panerai" model, a watch for Italian naval divers during World War II.

famous was: first one of six watches in a sales box was marked "Rolex", then two, and years later, three: a "half success" that took almost twenty years. Then in 1925 Wilsdorf decided: "Advertising must do it!" He invested a hundred thousand francs that year, mainly for advertisements in English newspapers that would make the Rolex brand stand for a special concept. At this time, though, there was already a new firm in existence, the present-day "Montres Rolex S. A."

The founding of this firm was, to some extent, done out of necessity, which had been brought on by a hard blow to the economy at the outbreak of World War I. In 1915 the British government had decided to levy an import duty which amounted to a third of the value of a watch.

Wilsdorf, who had imported his watches from Switzerland up to this point, exporting them then to all the world from London, went back to the land of his sources and soon transferred his export business to his Biel office that had been opened three years before. In 1919 he founded the aforementioned "Montres Rolex S. A." in Geneva, and divided the business in terms of organization and location. In Biel the movements were still produced by the "Manufacture des Montres Rolex, Aegler S. A." It was the job of the Geneva branch of the firm to create "models for cases acceptable to cultivated taste", finish the watches and market them. The movements delivered from Biel were first given a thorough examination in Geneva. After that they were set in operation and tested for accuracy for a week. After that they were put on sale.

Hans Wilsdorf set himself and his employees

198 ROLEX

Two "Oyster Perpetual" chronometers from the Forties, called "Hooded Bubble-Back" because of their covered band attachments and strongly arched bottoms, plus an "Oyster-Chronograph" with 30-minute and 12-hour counters, circa 1950.

five conditions for Rolex watches which have remained valid to this day:
1. create watch models that are popular among women as well as men,
2. develop a whole series of movements in a wide variety of sizes,
3. attain such a high quality that all calibers could be recognized as "chronometers" by the observatories.
4. make the results of their experience with watches that were made very particularly to attain the highest precision useful to current manufacturing, and
5. preserve precision, once achieved, by reliable and lasting protection of the movement from dirt and dampness.

Rolex watches had fulfilled the first four conditions very quickly. Wilsdorf's condition number five was the problem. It could only be solved if the movement was protected once and for all from various outside influences. The "complete precision" that the Rolex director demanded could be achieved only with a hermetically sealed case that would keep dust, sweat and moisture away from the precision movement forever.

In 1926 it was finally achieved. The Rolex "Oyster", the first fully watertight watch in the world, was invented. A year later it won the fame of Rolex that has lasted to this day. The great test took place on the arm of the London stenographer Mercedes Gleitze, who swam the English Channel in fifteen hours and fifteen minutes on October 7, 1927 and showed the astounded reporters the wristwatch that was still running.

Wilsdorf made commercial use of this success in several ways. On November 27 of the same year he had an advertisement printed on the title page of the "Daily Mail". That cost 40,000 Swiss francs, to be sure, but it made the watertight watch famous overnight.

Hans Wilsdorf also announced that in the future all watches from his house would have the name "Rolex" on the dial, the inside of the case and the movement. He also founded a system of his own representatives. In every English town there was a jeweler, several in larger cities, who had the monopoly for the sale of Rolex watches. The idea of displaying the Rolex "Oyster" in show windows also came from Hans Wilsdorf. In the middle of the display, in an aquarium with plants and goldfish, was a watertight "Oyster" watch that showed the astonished public the right time: an unusual method of advertising an unusual watch. It was justified, for the technical developments that are still included in every "Oyster" were and are unique.

In any case, the Rolex "Oyster" revolutionized the Swiss watch industry in 1927. After

Gold "Oyster-Perpetual" chronometer from the Forties. Two gold men's wristwatches with automatic winding, complete simple calendar and moon-phase indication, circa 1950.

Wilsdorf had spent hundreds of thousands of francs on advertising to make this invention known, the other manufacturers had to follow suit and offer watertight watches too.

Requirements for complete watertightness were:
1. a glass that was completely sealed,
2. a hermetically sealed case, and
3. a sealed winding crown.

For the first problem, Rolex invented a completely new material. Instead of glass they used an unbreakable synthetic material that would fit into the case precisely to fractions of a millimeter; it was produced in one of the firm's own factories. On September 21, 1926, a patent for the new case could be submitted, solving the second problem. It consisted of a central part, into which the movement was placed, with a metal ring. The upper and lower parts of the ring had a precisely cut threading. Thus the rim of the glass and the bottom were screwed against each other, absolutely parallel. The case was sealed—except for the winding crown.

But on October 18, 1926, the Rolex technicians had this problem under control too. The patent was submitted. The main problem here was undoubtedly that of sealing the winding stem hermetically and yet allowing the watch to be wound and set. Dust and microscopically small drops of water can obviously get into even watertight watches if, for example, a change of temperature causes low pressure in the watch case, which will actually suck impurities in. Wilsdorf created a seal with a spring in the crown that fulfilled all demands and, like a submarine periscope, provided an absolutely impervious case. When the crown was screwed upward, the spring caused the winding stem to connect with the crown, making winding and setting possible. When the crown was screwed down, this connection was released. Two smooth metal surfaces met. It could not be made any tighter.

With the invention of the watertight watch, Rolex had certainly won a leading position in the Swiss watch industry. But Wilsdorf saw a logcal consequence of the "Oyster": the creation of an automatic watch whose movement always wound itself, thus making the crown quite unnecessary. This was a problem that had constantly kept the watchmakers of earlier times busy too. Rolex solved it in 1931, after two years of research work, with the creation of the "Perpetual" and its legendary rotor, invented by Hans Wilsdorf himself. While the swinging weight of the automatic watches produced up to then moved back and forth between buffer springs, the centrally located rotor developed by Rolex could move freely in a circle. The semicircular turning weight will always take the lowest position it

Turned-up "Prince Brancard" model from the Thirties, in a striped yellow-white gold case. The name "Brancard" means "stretcher".

can, on account of gravity, when it has been set in motion, no matter how slightly. The automatic winding system was not integrated directly into the movement, but housed in its own frame and only connected to the barrel by a flange. A special provision prevented any overwinding of the mainspring. This all functioned soundlessly and almost without friction.

Hans Wilsdorf, who saw the "crowning of his achievements" in the "Perpetual", had this watch manufactured in three sizes, so that the desired model could be offered to women as well as men.

Then on September 21, 1935, out of a clear blue sky, a catastrophe struck the striving business that up to then had shown only climbing production and sales figures: the English pound was devalued. Hans Wilsdorf wrote in his memoirs:

"Our prices were known everywhere in English currency, and when we had to adjust to the new value, our exports fell to a third of what they had been before." . . . "It took our greatest efforts to open additional sales areas outside the British Empire, to which we had formerly limited our activity." . . . "At that time we opened a branch in Paris in order to get closer to the French market; in the same way we opened in Buenos Aires to enter the Argentine market. At the same time we took up our activity on a new basis in Italy, where we opened a studio for repair work in Milan as a technical support." . . . "We also decided to visit all the countries of Latin America, the Antilles and the Far East, especially China and Japan." . . . "In short, we tried everything and did not shirk any trouble."

But the best advertisements were doubtless the precision records of Rolex watches. Until 1939 it was Rolex watches that achieved the best regularity results in the whole watch industry at the Kew Observatory, in all sizes from the smallest movement to the "Prince" caliber. World War II made testing in England impossible. In place of that, Rolex scored 9.65 points in Neuenburg in 1942, the best result ever achieved to that time with a movement in the thirty-millimeter class.

The year of 1945 brought three decisive events for Rolex:
1. Rolex received, as was already mentioned, the fifty-thousandth running certificate for chronometer wristwatches from the official watch testing bureau in Biel—a result unique in the watch industry.
2. Rolex introduced the first watertight, automatic wrist chronometer in the world with self-acting date indication, enlarged two and a half times, through a window in the dial: the "Datejust".
3. After the unexpected death of his wife in 1944, the childless Hans Wilsdorf wanted to make sure that his business could survive after his death. He had all his shares gathered into the Hans Wilsdorf Foundation.

This foundation is unique in the Swiss watch industry. It is directed and watched over by the nominated members of the oversight council and bears the responsibility for the various Rolex companies in Geneva. The profits are divided according to exact specifications, which are explicated in the "Hans Wilsdorf" publication of the House of Rolex:

"A great portion of these funds goes, for

"Prince" model with springing digital hour indication at the sixty-minute position, case with stepped sides; Thirties. wrist.23

example, to welfare projects (in memory of the founder's wife), another portion to such trade institutions as, for example, the Watchmaking School in Geneva, the School for Fine Arts in Geneva (Industrial Arts Department), the School of Economic and Social Sciences at the University of Geneva, and the Swiss Watch Research Laboratory." The funds are meant above all to further technical research. And striking successes can be shown as results:

— A library for the blind was established, with recordings of literary and theatrical works.
— A laboratory for the Watchmaking School in Geneva was founded and equipped.
— The construction of a national exhibition pavilion for the protection of animals was financed in Lucerne.
— Film and radio equipment was made available for charitable purposes, and prizes are regularly given for the trade schools in Geneva.

In 1951, when Hans Wilsdorf was able to celebrate his seventieth birthday, fifty years in the watch business, 25th anniversary of the Rolex "Oyster" and 20th anniversary of the "Perpetual", a fund for pensions and social assistance, which is really exemplary in its generosity, was founded.

Life and Rolex success went on. In 1952 Hans Wilsdorf married again and, despite arthritis, traveled constantly. In 1953 he required surgery, but returned to his office and conscientiously continued to direct the life of Rolex. In the meantime the company had expanded to five buildings in Geneva and employed 750 workers. At the same time, 450 workers were employed at the sister company in Biel.

The further success of the brand can be seen clearly. In 1954 the first Rolex "Oyster-Perpetual" chronometer for women was produced. In the same year the firm surprised the market with a Rolex "Oyster Perpetual" chronometer with date indication, intended especially for pilots, while the "GMT-Master" allows the exact times of two time zones to be read. Two years later a Rolex classic appeared: the "Day-Date". It was the first wrist chronometer in the world with indication of the date and the fully spelled-out name of the day (now available in twenty-six languages). The last success that Hans Wilsdorf experienced before his death on July 6, 1960, was the diving trip of a Rolex "Oyster" (a special version of the case with a movement out of current production), the original of which is displayed in Washington today, while a copy is shown in Geneva with no less pride. The watch, which was attached to the outside of the diving ship "Trieste" of Professor Piccard, dived into the Mariana Trench, near Guam in the Pacific Ocean, to a depth of 10,916 meters, representing a pressure of nearly a ton per square centimeter. Afterward it ran (naturally) perfectly.

Four more dates after the death of Hans Wilsdorf are still stressed today:

— In 1971 the Rolex "Oyster Perpetual Sea-Dweller 2000" appeared, a diver's watch guaranteed watertight to 610 meters (2000 feet). It is the first diver's watch with a helium vent, for at great depths helium passes through the glass into the case. The diver's

Wristwatch with chronograph and complete simple calendar, from the Forties.

The development of the Rolex "Oyster" models from 1926 to 1955. The cushion-shaped model patented in 1926 has a hand-wound movement, the others are equipped with automatic winding.

204 ROLEX

1959   1970   1980

The development of the Rolex "Oyster" models from 1959 to 1980. The crown is protected against damage, the turning bezel can be set for diving time only counterclockwise for reasons of safety.

decompression stages on the way up are too short for the watch; without a helium vent the watch would explode during decompression.

— In 1978 Reinhold Messner wore an "Oyster-quartz Wrist Chronometer" on his arm during his historic climb of Mount Everest without an auxiliary oxygen supply.

— In 1980 Rolex introduced the "Sea-Dweller 4000", which is watertight to 1220 meters (4000 feet) and equipped with a helium vent and sapphire glass. Its dependable chronometer movement is fitted with a mechanism for quick correction of the date.

— April 1985: since 1961, more than 4.1 million Rolex movements have borne the official title of "Chronometer". Rolex is thus by far the greatest Swiss producer of chronometers. Yet these watches make up only about 1% of the total Swiss watch production.

Hans Wilsdorf did not live to see the move of Rolex into a new building on Rue François-Dussaud in Geneva. The modern complex of buildings (with security provisions including infrared radar alarms, television monitors and control computers) is surrounded by 600,000 liters of water. This modern castle moat symbolizes the "Oyster" model that can always be surrounded by water without being harmed. The new building also houses the firm's unique watch collection, which was begun in 1940 by the founder of Rolex, the "Hans Wilsdorf Collection" with 140 watches and enamel works from the 17th, 18th and 19th centuries (value: over a million Swiss francs), with a gold watch by Breguet ("Perpetuelle") which shows the day, date and moon phases, has minute striking, shock-resistance and winding indication. The collection can be viewed with written permission (Montres Rolex S.A., CH-1211 Geneva 24).

The construction of the new building simultaneously made it clear that Rolex no longer rented quarters in its native land, just as all its worldwide branches are housed in buildings owned by the firm—property worth millions of Swiss francs. Just a few examples:

— New York, the Rolex Building on Fifth Avenue,
— Bombay, Mahatma Gandhi Road,
— Paris, Aevnue de la Grande-Armée,
— Cologne, Bahnhofstrasse 1-9.

In all, Rolex is represented in 24 of the world's great cities: Bangkok, Bombay, Brussels, Buenos Aires, Caracas, Cologne, Dallas, Djakarta, Geneva, Hong Kong, Johannesburg, London, Madrid, Manila, Melbourne, Mexico City, Milan, New York, Paris, Sao Paulo, Singapore, Taipei, Tokyo and Toronto. In Geneva the only regret is that there is no Rolex branch in space! The shares of the market taken by men's and women's watches is regarded by the Rolex directors as "balanced". The world sales of their watches "quarter" the globe with 25% each for the USA, Europe, Asia, and a combination of the

The octagonal Rolex "Oyster" that Mercedes Gleitze wore on her arm while swimming the English Channel on October 21, 1927. The bottom is engraved as a memento of that day.

Middle East, Africa, South America and Australia.

To be sure, this balanced division came about mainly in the last ten years, in which the Middle East's share of the market decreased considerably.

For Rolex time is "rhythm, tempo, harmony, measure and movement" (the firm's motto), but behind it all, time is also money, for whoever produces and sells about 500,000 chronometers a year can also say of himself with necessary understatement that Rolex does not need banks, but the banks work with Rolex's money.

Whoever wants to know what makes a watch a Rolex should have a look behind the brown walls and see their origins in the air-conditioned studio. Since the "Rolex-makers" (there are about a hundred of them) scarcely move when they, for example, assemble an "Oyster" out of approximately 220 individual parts, the temperature in the production rooms, with natural light from three sides and the atmosphere of a dental laboratory, is kept by the air conditioning at three degrees above that in the business offices. The parts that they assemble are received from the sister firm in Biel or other suppliers, such as the chronograph movements from their SMH daughter Valjoux, or the movements of the current Tudor models from Eta in Grenchen. Bands and dials also come from specialist firms.

Along with modern tools, one further achievement of our days has found a home in the studio. This is the Walkman, that is so popular and so widely used that one could believe Sony had invented this cassette recorder with earphones just for the workers in the Swiss watch industry. Very few employees wear a Rolex on their wrist, though after working for the firm for a year they can buy for half price and their families get a 40% discount. The public relations department, which serves dealers, Rolex owners and fans, is supplied with "watches on loan". For these employees, a Rolex is part of their work attire and, unless it is bought in used condition, must be returned if they change their workplace.

Rolex owners are always welcome at the factory, for information about watches that change hands is reported in writing only in exceptional cases. If someone has an antique Rolex with problems, he should come to Geneva. All problems will be solved there, it is promised. If the visitor wishes, he will be shown how a watch is made.

The case of an "Oyster", which takes a year to assemble and is watertight to a depth of 100 meters, is worked from a massive block of stainless steel, gold or platinum. 163 work processes are needed before this piece is finished and can be sent to Geneva. The manufacture of a "Twinlock" winding crown requires thirty-two work processes. Now there is also a "Triplock" crown, which was developed for divers' models (300- or 1220-meter depths and can be recognized by three features). The finished hand-polished case with glass and bottom is given a first water-tightness test in Geneva. It goes into a Mariotte meter, which uses electronics and air pressure to make any lack of tightness show up at once. In such a situation, the case is immersed in water in a vacuum measuring device, and the

A golden "Oyster" case and cut model of the screwed-in winding and hand-setting crown in the "Oyster" case.

non-tight place can be seen by means of air bubbles without letting water into the case. The error is corrected, or else the case is rejected. Every "Oyster" must pass this test again after the installation of the movement and before it is shipped out. The watertightness test is demonstrated to visitors on specially developed models that are smaller and more attractive but function by the same principles. The watch crystal is made of scratchproof synthetic sapphire glass (3 mm thick for the "Sea Dweller 4000") or unbreakable plastic. The glass is 100% watertight; the main job of the bezel is decoration. The precisely cut threading of the case bottom itself is enough to guarantee watertightness, but Rolex adds four thickening rings made of a secret material. The case is screwed together by special machines that make sure that it can be opened only by Rolex specialists. The finished watches are given a thorough test of running precision. A machine tests the watch photomechanically in three positions and compares it to an atomic clock. Only watches that deviate by no more than minus one or plus five seconds per day are sent to an independent institute for chronometer testing; the others are dismantled and rebuilt. To be allowed to bear the official title of "Chronometer", each individual Rolex must prove its precision for fifteen days and nights. Only then does it receieve the official certificate, of which Rolex earns 86% of all those given in Switzerland. Before the watches are shipped out, the serial numbers, a consecutive number in connection with the official chronometer number, and the Rolex branch that will receive it are noted.

At this time, in addition to the usual

The Rolex headquarters in Geneva.

The massive gold blocks from which the "Oyster" cases are made.

On the outer wall of Jacques Piccard's Bathyscape, this special edition of a Rolex "Oyster" dived to a depth of 10,916 meters, at which an external pressure of about 1000 kp/cm³ was attained.

mechanical "Oyster" models in stainless steel, gold or platinum, with a variety of cases (special orders are possible), quartz watches are also produced in the "Oyster" and "Cellini" lines.

Special types of "Oyster" models are:
— the "Day-Date", which combines all the special features invented or refined by Rolex: the watertight "Oyster" case, the self-winding system with a rotor, the officially tested chronometer movement, the indication of date and fully spelled-out weekday. This perfect watch is available only in 18-karat yellow gold, white gold or platinum. It may be noted in this context that Rolex has the highest industrial gold consumption in Switzerland;
— the increased number of swings of the "Perpetual", namely 28,800 half-swings per second, as compared with 18,000 half-swings by the hand-wound models, an essential factor in their higher precision;
— the "Milgauss" model is equipped with a special anti-magnetic protection. In experiments it kept running undisturbed in magnetic fields of 5000 Oersteds, while normal watches stopped at 100 Oersteds;
— the "GMT Master" with simultaneous time indication for two time zones;
— the "Submariner Date", an automatic chronometer, watertight to a depth of 300 meters, in a stainless steel, steel-gold or 18-karat gold case with a diving-time setting ring that can be turned only counterclockwise, the "Triplock" winding crown protected against damage by metal shoulders, and the "Fliplock" watchband, by which the watch can also be worn over a diving suit;
— the "Explorer I and II": the "Explorer I" (#10160) was made for the British Everest expedition of 1953, which led, under the direction of Colonel John Hunt, to the first climbing of the world's highest mountain. These watches, that dealt with temperature changes of 70 degrees, were atypically equipped with leather watchbands on account of the cold at great heights. The "Explorer II" (#16550) has an additional red 24-hour indicator. It was developed especially for cave explorers who can see by the red hand and 24-hour bezel whether it is day or night when they are underground;
— the "Sea Dweller" with the patented Rolex helium vent that makes possible diving to depths of 1220 meters, and finally

From left to right: "Oyster", steel watchband.

"President", yellow gold watchband.

"Jubilé", yellow gold watchband.

View of a movement with rotor removed.

The "Oyster Perpetual Day-Date" model, with date and day indication, as well as a "President" armband.

210 ROLEX

View of the various "Oyster" models, beginning with the "Oyster Precision" with hand winding, ending with the "Oysterquartz" with electronic quartz movement.

—the "Oysterquartz" with long-lasting battery and a quartz frequency of 32,768 swings per second. A deviation of less than one minute per year is guaranteed. The watch is watertight to 100 meters and resists a magnetic field of up to 1000 Oersteds. Rolex developed its own measuring devices to test its precision.

In order to inform potential customers and Rolex owners of the quality of their watches, Rolex spends about sixty million Swiss francs a year for marketing and advertising. Well-known personalities (such as violinist Yehudi Menuhin, golfer Severiano Ballestreros, conductor Antal Dorati, mountaineer Reinhold Messner, director Franco Zeffirelli, racing driver Jackie Stewart, author Frederick Forsyth) are put under contract for advertising after they have already worn a Rolex for one to two years. At least this is the normal situation, it is stated in Geneva, since product placement is forbidden as being in bad taste. They speak of legal difficulties in Germany, where it is not mentioned that the stars of the American TV series, "Dallas", were given Rolex watches, or that in the film "Marathon Man", Dustin Hoffman, while chased by gangsters, exchanges his watch with the golden crown for change to make a telephone call.

Advertising with well-known personalities in public life was begun quite early in Geneva. After Winston Churchill received a chronometer engraved with his arms in the Forties as a gift from his admirer Wilsdorf, and later General Eisenhower wore a Rolex when he became President of the USA and moved into the White House (whereupon Rolex stressed the fact that "he did not receive it from us as a gift"), and when US Secretary of State John Foster Dulles was seen with a Rolex on his wrist, the advertising specialists created the slogan: "Men who guide the destinies of the

Three wristwatch models from the "Cellini" line, one "Oyster Perpetual Datejust" in steel-gold ("Rolesor") and one in gold.

ROLEX 213

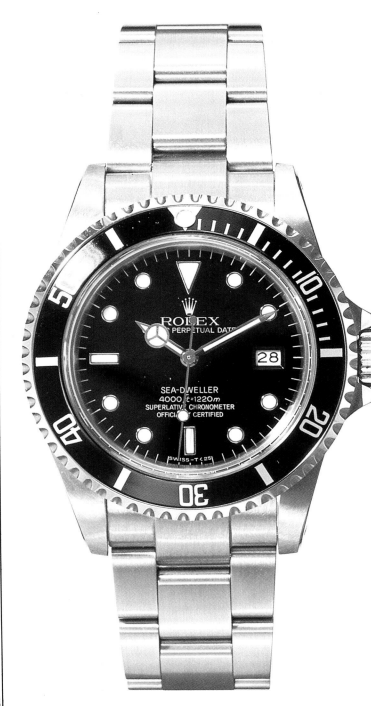

world wear Rolex watches". During the Fifties and Sixties, Adenauer, de Gaulle, Nixon, the Shah, Qaddafi and Castro fit into this campaign. They all wore, and still wear, Rolex watches. Their photos and portraits are secretly preserved today in the "Hall of Fame" behind closed doors on the main floor.

Counterfeit watches are no longer a problem today. Quiet prevails in this realm, after several million Swiss francs and many courtroom victories brought an end to this abuse.

The success of Rolex rests on the fact that the firm essentially builds chronometers of a model and leaves "playing games" to others. And the future of the brand is seen thus in Geneva:

Concentration on a lasting piece of jewelry that tells the time and gives prestige. Production increases beyond the present 500,000 pieces per year, with an income of over a billion Swiss francs and handsome profits, would change the character of the product and are not planned. Rolex at least does not want to "grow rapidly".

In fact, Rolex has found its place on the world market and wants to hold it qualitatively. The mechanical series watch will be developed further. As for quartz watches, the firm is waiting and tending toward homogeneous quality and long life. The "weak spot" battery, in any case, is to be eliminated. Why, the managers ask, should they not develop the rotor of the "Perpetual" as a generator (like the light machine that charges an automobile battery)? In this spirit Rolex research will proceed equally in the mechanical and electronic areas alike.

"Oyster Perpetual Datejust" for women. The bezel is decorated with sapphires and diamonds, the dial with diamonds. "Jubilé" band.

The "Sea Dweller" model with automatic winding and date indication. "Oyster" case, watertight to 1220 meters. The side view shows the helium vent.

ULYSSE NARDIN

It was 1774 when the twenty-year-old French artisan Jean Léonard Nardin packed his worldly goods and tools on a horse-drawn wagon, cracked his whip and set out for Switzerland. In Le Locle, high up in the Swiss Jura, he found a new home and job. "Master Jean", as everyone in the little town of blacksmiths and weapon-makers soon called him, earned his living by building stoves and small water systems, some of which are still in operation, which really speaks well for his concept of quality which has been inherited by succeeding generations.

In 1792 there was born to him a son who was also interested in technical things, though

mainly in small-scale precision. Léonard Frédéric Nardin became the family's first watchmaker. But only the grandson of "Master Jean" was to give his name in 1846, at the age of twenty-three, to a new brand of Swiss watches: Ulysse Nardin. His name was soon to become known to all the world, for the young watchmaker proved his love for accurate work as well as his talent as an artisan by the precision of his timepieces. He expected the same from his suppliers, and a list of them today gives us a collection of the best watchmakers in Switzerland at that time.

In 1862 his mechanical works of art gained the highest award, "The Price Medal", at the World's Fair in London. Thus the world stood open for Ulysse Nardin. His complicated watches, particularly pocket watches, found friends in America too. Gold medals, Grands Prix and other awards followed, one after another, and in the course of time brought the Swiss firm the greatest number of awards in the watch industry at that time. They attested to Nardin's striving for sensationally high quality in his products. One example is the watch that was built as a Swiss contribution to the World's Fair in Chicago (1893). A team of experts estimated that if one wanted to build this watch today, 700 work hours would be needed just for the case. In fact, duplicating it without a casting of the original would be as good as impossible.

It was above all one particular masterpiece that brought the firm its worldwide fame: the marine chronometer that won 4300 awards from observatories. In the days before quartz technology or time broadcasting by navigational satellites, the exact position of a ship

Left: Jean Léonard Nardin

Right: Ulysse Nardin

The firm's headquarters in Le Locle Document from 1966 bestowing a prize on Ulysse Nardin for a series of officially tested marine chronometers.

could be determined only with sextants and chronometers (two of each were always on hand to be safe). Good shipping lines ordered their chronometers from Ulysse Nardin, who, as the awards from the observatories showed, produced the most precise chronometers in the world.

But in the past few decades, technical developments made the marine chronometer quite superfluous. And the descendants of Ulysse Nardin were, as lawyers, too far away from their inherited business to understand the demands of new marketing correctly. Using designs that offered nothing new, but a lot of gold on the watch, they tried to conquer the Near East market in the Seventies—in vain. The warehouse on the Rue de Jardin in Le Locle grew fuller and fuller, the network of dealers went to pieces, and the debts increased.

Then in 1982 Rolf W. Schnyder's day dawned. For approximately one and a half million Swiss francs he acquired a 60% majority (the total stock capital was one million francs), the debts, the lack of retailers, and above all, the problem of how to catapult Ulysse Nardin back into the leading group of Swiss manufacturers. He had the solution in mind, theoretically: Ulysse Nardin had to present a watch that did not yet exist. But what kind of a watch could that be?

Rolf Schnyder was, to be sure, no stranger to the watch business, but the entrepreneur who had produced dials, cases and watchbands for Swiss manufacturers in Malaysia with a work force of some six hundred did not find the answer with ease. While he had most of the unsellable models of his predecessors melted down, built up a small network of retailers for the new mechanical and extremely precise wristwatches, his thoughts centered on the unique wristwatch of the future. It was to carry the fame of Ulysse Nardin to the whole world again.

During a visit to the studio of Jörg Spöring in Lucerne he looked for new ideas and almost stumbled over the solution. In the reception room he saw an astrolabe, about a meter high,

One-minute tourbillon chronometer-regulator with separate hour and minute indications. Individually prepared for the house collection of Ulysse Nardin in 1987.

Back of the wristwatch with tourbillon, with a view of the movement. All parts are polished and smoothed by hand.

Gold wristwatch with chronograph, 30-minute and 12-hour registers, complete simple calendar and moon-phase indication, watertight to 30 meters.

The famous "Astrolabium Galileo Galilei" with automatic winding and many astronomical indications, first displayed publicly at the Basel Clock Fair in 1984. The Guinness Book of Records honored this watch by depicting it on the cover of their 1988 edition.

that showed not only the time but the positions of the stars. He asked who had built it, and Spöring told of an old commission from the Vatican, which had the Farnesian Clock restored and repaired in 1978. The work had been done by Ludwig Oechslin, a truly remarkable universal genius.

Before he became a watchmaker, Oechslin had studied archaeology, ancient history and Greek, only then discovering his preference for handwork and mechanical watches and beginning his studies in a Lucerne studio. And Oechslin had traveled to the Vatican and begun his work, but in order to make progress, he had had to study astronomy, physics and mathematics, finishing his work plus further study of physics, philosophy and the history of science with a doctoral dissertation on the clock as a model of the cosmos.

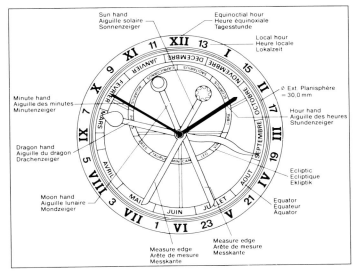

The astrolabe was a "by-product" over which Rolf Schnyder "stumbled" while Dr. Oechslin developed perpetual calendars for pocket watches.

He wanted to make such an astrolabe as a wristwatch, the new manufacturer said to the learned fine mechanic. The latter nodded and said it must be possible. A year and a half later the prototype, which now lies in a vault in Le Locle, was finished. But there were problems that stood in the way of producing even the smallest series, for the movement was simply too thick for a wristwatch. Rolf Schnyder called in the constructor Urs Gyger, who had created the "Eterna-matic", once the thinnest automatic wristwatch in the world. Gyger and Dr. Oechslin improved the mechanics together and made inventions where no progress had been possible with existing equipment. The epicycloid miniature mechanism with four simultaneous indications was patented, as was the new space-saving ball-bearing system for ensuring the stability of the mechanism. By the beginning of 1987, 150 of these watches had been built and sold, representing a production of fewer than a hundred examples per year.

Museums and collectors were the first purchasers of these automatic wristwatches, whose mechanics Dr. Oechslin had worked out so well with a pocket calculator that, according to computer calculations, there would be a one-day deviation from the actual position of the stars only after 144,000 years. The Moroccan royal family ordered four "Astrolabium Galileo Galilei" with a special Arabic engraving; Henry Kissinger, Secretary of State of the USA from 1973 to 1976, also owns one. Rolf Schnyder did, though, have to rewrite the

Explanation of the various astronomical indications of the "Astrolabium Galileo Galilei".

Man's wristwatch with split second chronograph and 30-minute register, from the Twenties.

directions three times to give the owner understandable instructions for its use. For it is not simple, "to wear some of what Galileo Galilei had in his head on your wrist" (advertising slogan). One must first have understood the system of the stars and of time measurement in order to be able to understand this masterpiece of the watchmaker's art, which simultaneously indicates the sun time, star time, month, zodiac, lengths of day, night and twilight, the moon phases, the astronomical coordinates of sun and moon and the aspects of the two to each other (sun and moon darkness). What an amateur astronomer would have to go to the trouble of working out from tables, this watch shows the trained eye at a glance.

Then, of course, one can predict, as did Rolf Schnyder to an American while vacationing in the Alps in broad daylight, that the moon would rise in half an hour "about over there, quite high above the horizon". The American tourist laughed at first, but was so amazed in half an hour that on the same evening he ordered "this fantastic watch".

The heart of the watch is a patented drive, in which the movement turns around itself once every twenty-four hours via the winding mechanism.

Man's gold chased skeleton wristwatch with automatic winding.

Man's gold wristwatch with chronograph, 30-minute and 12-hour registers, date and moon-phase indication.

For this, single toothed wheels were doubled in the drive train, four different indications are arranged on one arbor, a special light-metal alloy guarantees shock resistance—all that in nine millimeters of height. This achievement did not just fascinate the world of watchmakers. The British Museum displays this watch, which is supplied on request with a mahogany box in which the watch turns when the owner has taken it off, so that the indications can be read even when it is not being worn. In addition, this eighteen-karat gold work of art is watertight to a depth of thirty meters and has dials divided into five-degree steps, individually suited to the geographical location of its owner's home. The stars were right for Rolf W. Schnyder, who originated the factory's slogan, "beyond hours and minutes", and who produces, with his staff of sixteen, some thirty-five other models, all mechanical, in quantities from 2500 to 3500 pieces per year. Germany, Switzerland, Italy and England are his main markets; he is just getting a foothold in the USA. Thus the firm's own service center was opened in New York, at which the watchmakers are ready to equip the "Astrolabium" with the right dial and set the hands correctly.

Ulysse Nardin's newest creation is the "Michelangelo", which naturally has a mechanical movement (in an 18-karat gold or steel-gold case), with day, date, month and moon-phase indication. The greatest precision in its production guarantees that this rectangular wristwatch is watertight to a depth of thirty meters. The sapphire-glass bottom gives a free view of the movement and lets the owner convince himself of the uniqueness of his watch. This "Michelangelo" proved that the name of Ulysse Nardin sounds good again. The first advertisement (a third of a page in the weekly magazine "Der Spiegel") immediately elicited countless inquiries and eighty orders. The first year's production was sold out at once.

The proprietor wants to keep his staff small in the future too. His formula: many jobs are done in work at home by masters of their trade. An example is the watchmaker who masterfully chases the individual parts of the superb skeleton watches exclusively for Ulysse Nardin.

Something else of a spectacular nature is being prepared in the Rue de Jardin at present. Something that will make its mark on history like the Astrolabium. One can believe Rolf Schnyder when he makes this announcement, for he has ordered new machines and tools. They are being built in the attic of the factory—for Dr. Ludwig Oechslin.

"Michelangelo", a mechanical masterpiece of lasting value: an automatic 18k gold dial with mother-of-pearl inlays, arched sapphire glass. The movement is visible through the sapphire bottom.

Man's gold wristwatch with minute repetition, No. 9827, whose movement was finished in 1899 and later built into a wristwatch case.

VACHERON CONSTANTIN

The price was remarkably high, even for a luxury wristwatch, on the Rhône Island in the heart of the watch capital of Geneva they are no longer proud of the entry in the Guinness Book of Records. At the end of the Seventies the firm of Vacheron & Constantin introduced the most expensive wristwatch in the world, the "Kallista", Greek for "the most wonderful".

The watch deserved its name, for despite its price of five million dollars it was sold at once. (Today there is only a copy displayed in the conference room at the factory.) Case and band were chiseled out of a massive kilogram bar of gold. After being worked, the weight of the gold still amounted to 140 grams. The case and band were decorated with 118 emerald-cut diamonds, between 1.2 and 4 karats, each with its own certificate; only the hands can be seen. Neither the value of the gold nor the 6000 work hours that went into making this watch account for its unthinkable price, but rather the more than 130 karats of jewels.

From then on, the world's watch fans associated the name of Vacheron & Constantin with unattainable luxury. Meanwhile, people at the factory turned their thoughts back to the tradition of the old master watchmakers, to real handworking art, and advertised with the slogan, "A Vacheron & Constantin is not a matter of price", seeking customers who wanted a watch from one of Geneva's oldest manufacturers.

It was 1755. In that year the English under George II waged war against the French on land in America, India and Germany, and at sea all over the world. In Potsdam Frederick the Great dreamed of making Prussia a great power, Madame de Pompadour led France, under her royal lover Louis XV, toward its downfall, the world danced to Haydn's minuets, and 800 master watchmakers lived in Geneva and had joined together in the guild of the "Cabinotiers". Along with their handwork, which they worked hard to raise to as high a level as possible through progressive thinking, they also practiced the fine arts. And a master watchmaker who composed music that was performed was no rarity. The "Cabinotiers" were artisans, talented artists, scholars, philosophers and lovers of perfection all in one, an aristocracy of work, as the

JEAN-MARC VACHERON 1731 * 1805

FRANÇOIS CONSTANTIN 1788 1854

Socialist Internationale cynically put it a century later. At that time strong business interests may have been involved inconspicuously, for it was a privilege of the noble and rich to know what time it was. And they were at once the buyers of the exquisite timepieces and the patrons of the artists.

Among the aforementioned eight hundred Cabinotiers, who employed 4000 apprentices and journeymen, was Jean-Marc Vacheron, who had settled in the picturesque Saint Gervais district of Geneva in 1755. The twenty-four-year-old opened a shop along with an apprentice, built pocket watches, and it was his personal ambition to build timepieces that would stand out for their quality and particular elegance. But perfect handworking ability alone did not suffice to achieve his goal. A considerable amount of business sense was also needed. And he had that too, for he sold many of his pocket watches to italian and French nobles. A pocket watch from 1760, signed "J. M. Vacheron", exists today that is not only a perfectly produced piece of work but was formed and decorated very extravagantly.

In 1785 Jean-Marc Vacheron turned the leadership of the business over to his son Abraham, who directed it successfully. It was no easy task, for the French Revolution caused the nobility of what had once been the elegant kingdom of France to spend their time saving their lives instead of buying exquisite timepieces. After the proprietor's marriage the firm was called "Abraham Vacheron-Girod". From this era too a watch has survived, signed with that company name and numbered 33, 498.

The firm lived through all the upheavals and, from 1810 on, was directed by the third generation, Jacques Barthelemy Vacheron, now under the name of "Vacheron-Chossat" on account of marriage. Under the most severe conditions, as a letter to his father indicates: "You ask me how things are in Geneva... The workers have never been in such a bad situation... Most shops are closed."

Yet the House of Vacheron, which had supplied the most elegant salons of Europe in the Eighteenth Century, survived that terrible time. As early as 1814 the economic situation improved, and the firm produced not only pocket watches but also music boxes with two

Man's gold wristwatch with date indication and sweep second; quartz movement.

tunes, repeating movements, thin movements and cases with smooth, machined or diamond-set finish. With these wares, Jacques Barthelemy Vacheron traveled to France, Germany and Italy three times a year, especially to Milan, Venice and Trieste. These long trips brought the company many orders for costly pieces set with jewels, repeating watches or small music boxes.

Despite all his success, Jacques Barthelemy Vacheron slowly began to realize that the business could not be managed indefinitely from afar. Thus he decided in 1819 to take his friend François Constantin, the son of a rich textile and grain merchant, into the business as a partner. From then on the firm went under the name of "Vacheron & Constantin". François Constantin could show experience in this kind of business, for he had worked as a salesman for the respected watch company of J. F. Bautte. And that was exactly what his job was to be at Vacheron & Constantin.

In any case, he was a capable businessman, as the further upward development of the firm indicates. At first, though, the renowned watchmakers of Geneva muttered about the

Men's gold wristwatches with complete simple calendar and moon-phase indication, hand-wound movements; Forties and Fifties.

"decline of good manners" and predicted that François Constantin would be more likely to lead the firm to ruin than to success. This view was based on the extravagant habits of the new partner, who did not behave at all as modestly as was proper for a watchmaker. Wherever François Constantin turned up, he was not concerned about a few francs. One must have had the impression that the "King of the Watchmakers" was holding court, opulently and excessively self-confidently, flaunting his role as the representative of an international business house. François Constantin lived up to the slogan that was to accompany the firm through its next decades: "Faire mieux si possible, ce qui est toujours possible." (Do better when possible, and it is always possible!)

At the factory the jobs were divided precisely: Abraham Vacheron, now sixty years old, was in charge of quality control. His son Jacques Barthelemy Vacheron was occupied almost exclusively with directing the business in Geneva. The technical status and capability of the factory were suited to the manifold wishes of the clientele.

Left outside and center: men's round gold wristwatches in imaginative cases, hand-wound movements; the model came on the market in 1952.

Right center: man's gold wristwatch with hand-wound movement, circa 1950.

Right outside: man's wristwatch in steel-gold, Lepine caliber with the winding crown at the 12; Thirties.

Lower right: Man's wristwatch in art-deco style; platinum case with gold and enamel inlays; Twenties.

But products of a so-called second quality were also sold, though it is not recorded as to how the division was made. This second quality was not signed "Vacheron & Constantin". The leading models of the second class were engraved "Abm. Vacheron à Genève" or "Chossat & Cie.", the others "Abraham Vacheron".

François Constantin was the born salesman, which was not as easy as it sounds today. One must imagine oneself in those times to be able to evaluate the 31-year-old Constantin properly. The German duchies were shaken by an economic crisis, Louis XVIII reigned in France, tariff conditions in various lands were often changed arbitrarily overnight, valuable watches damaged or destroyed during inspection, fines imposed out of nowhere. Payments in arrears and "kite-flying" speculation were prevalent too, and many a nobleman would not admit having placed an order when François Constantin stood before him with the finished watch.

In 1824 François Constantin set out for Italy. In Livorno he traded several watches for a number of barrels of wine, since the dealer wanted watches but had no cash.

Constantin arranged to have the barrels shipped to Geneva and sent his partner precise written instructions with them.

"It will be best to transfer the wine into small bottles. Then advertise them in the "Feuille d'Avis", naturally selling at a high price. But also prepare to sell them at a lower price if someone wants a goodly quantity of them. In any case, do your best to make a profit."

Lower left: gold wristwatch with center second; Fifties (see also page 223).

Center left: man's wristwatch with the crown by the 12, Thirties (see also page 223).

Center and outside right: Men's wristwatches with chronograph, 30-minute register and tachometer scale; Forties and Fifties.

On July 18, 1955, Messrs. Bulganin, Eden, Eisenhower and Faure met at the "Palace of Nations" in Geneva to discuss questions of world peace. On this occasion each of the statesmen was given a Vacheron & Constantin watch by twenty citizens of Geneva, with the inscription: "May this watch always show happy hours—for you yourself, your people and the peace of the world."

Vacheron & Constantin protected themseles against highway robbers in an original way: the expensive watches were packed for travel in huge cabinets that were so big that they could not be transported without special preparations. Complex locks and secret compartments also made it impossible to steal these "treasures". Along with this wealth of originality, François Constantin also had a fine sense of political developments. In 1821, when the situation in the profitable sales areas of Italy and Austria became increasingly worse, he decided to open markets for Vacheron & Constantin in other lands. Under the prevailing conditions, that was a difficult undertaking. But by 1830 the first watches were being shipped across the Atlantic toward North America. At first they were sold by Jean Magnin in New York, soon afterward by Brey in New Orleans, and in 1864 the firm's own agency was opened.

Vacheron & Constantin had introduced a real revolution in industrial watch production in 1839. Because of growing demand, it had to be considered how to increase production without having a deleterious effect on quality. Jacques Barthelemy Vacheron remembered a certain Georges-Auguste Leschot. In the process of his watchmaking training with his father, Leschot had carried on remarkable experiments with machines that could be used for series production of watch components. At that time the very idea shocked the watchmakers of Geneva. To be sure, Leschot had also invented an improved lever escapement, but developing machines for series production of watch components brought him nothing but scorn and ridicule at first—until on one June day in 1839 Georges-Auguste Leschot walked into the factory and was immediately hired as technical director of Vacheron & Constantin, the firm having contracted to assume all the costs of developing the machines made by Leschot. Vacheron wrote to his business partner:

"We are in the process of completely changing our production methods. Our precision watches will soon be scarcely more expensive than other, lower-quality products."

Outer left: man's square wristwatch from the Fifties.

Inner left and right: men's wristwatches from the Forties.

Outer right: man's rectangular wristwatch from the Thirties.

Leschot went to work. He built tools to build his machines, worked day and night, and two years later, in 1841, he presented the results of his creativity:
- A type of cutting machine that worked according to the principle of the "stork's-bill" (pantograph). One pushed large levers and produced several small identical watch parts simultaneously. Fast, simple, exact, and making more pieces than before.
- A turning and drilling machine that could be set up to drill and sink holes, always in exactly the same position.
- Other turning and cutting machines, used for, among other things, the working of bridges and blocks, plates, the teeth of escape wheels, and the cutting out the insides of the escape wheels.

The factory had now become an industrial concern. The watchmakers, who had formerly feared for their livelihood, quickly made friends with the machines. At last they had more time to use their artistic and handworking capabilities for working the pieces.

The death of Abraham Vacheron in 1833 led to some changes in the company. In 1844, after moving to the historic "Tour d'Ile" between the two Rhône bridges, Jacques Barthelemy Vacheron turned the leadership of the business over to his son Cesar, who had been carefully trained for the job. He also published the essential capabilities of manufacturing watch parts with Leschot machines.

In 1845 Leschot and the firm were awarded the "Auguste de la Rive" by the Societe des Arts. Thus even the harshest critics were convinced, and Vacheron & Constantin received many new orders. The quality of their thin calibers was so exceptional that most watchmaking firms built Vacheron & Constantin movements into their watches. New markets opened through agencies in the Dutch East Indies (1847) and Calcutta (1850) brought further increases.

Around 1850 Vacheron & Constantin decided to build crown winding into their watches. The first pocket watches thus equipped were developed, but at first they were not satisfactory. In 1855 the first watch of the type was sold. A year before that, the tireless "selling machine" François Constantin had died, and his job had been taken over by his 24-year-old nephew Jean-François Constantin—with equal success, as the further growth of the business shows. At that time the rich and powerful of the world called Vacheron & Constantin watches their own. In Paris, Empress Eugenie owned a pendant watch; a thousand kilometers away. Tsar Alexander II had a pocket watch. Emperor Napoleon III, Eugenie's husband, ordered several watches with his portrait and used them as presents to important personalities.

From 1860 on, the company's technical development had advanced so far that they

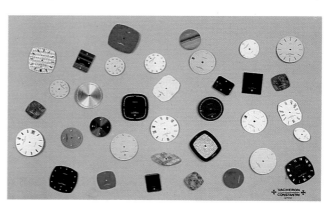

A selection of the rich spectrum of dials by Vacheron & Constantin.

Man's gold wristwatch with automatic winding and perpetual calendar, leap-year and moon-phase indication, on the market since 1983.

could now produce all individual parts of a watch themselves. Vacheron & Constantin could manufacture watches without depending on suppliers of parts, even though the goal of making the parts "interchangeable and ready to use" had not been achieved. Every part that came out of a machine had to be worked further by hand if the firm's own quality standards were to be maintained. So they made a virtue of necessity that is still practiced: pre-finishing by machine, then finishing by hand!

After the death of Jacques Barthelemy Vacheron, the firm decided in 1864 to stop selling raw movements to competitors. This move was meant to maintain a technical advantage. A new feature of the movements of that time was a lever escapement developed by Leschot, which was now machine-made too.

In 1865 the Emperor of China bought a pocket watch from Vacheron & Constantin, a significant sales success for Jean-François Constantin. Two years later, for unknown reasons, he stepped down from the company's

Man's wristwatch with automatic winding, perpetual calendar, leap-year and moon-phase indication, in skeleton form; platinum case.

View of the complex mechanism for the perpetual calendar.

management. No explanation can be found in the annals of the firm. It is known, though, that the 37-year-old continued to work as a "normal salesman" and Cesar Vacheron took over the sole direction of the company. The firm's name was changed again, becoming "Cesar Vacheron & Co." Very few watches have survived from that era, since it lasted only two years. Then Cesar Vacheron died at the age of only 56.

Charles Vacheron, 24 years old, took over the business, and the firm was now renamed "Charles Vacheron & Cie.". One year later a new stroke of fate struck the company, when its young proprietor died at age 25.

Now a woman assumed the direction of the firm, which caused a sensation in the Swiss watch industry. As if that was not enough, another woman assisted her. Things would not go well with women in charge, the Watchmakers' Guild of Geneva predicted. And many a watchmaker secretly regarded himself as the right man to take over the machines and the technical know-how. But the ladies proved that they could manage a watch manufacturing company thoroughly.

The two women, Laure Vacheron-Pernessin, the widow of Charles Vacheron, and Catherine-Etiennette Vacheron, the 88-year-old widow of Jacques Barthelemy Vacheron, made a series of smart decisions in the five years they worked together, during which Catherine-Etiennette never officially belonged to the firm's management. They laid the groundwork for the further development of the company into the Twentieth Century. At first the firm's name was changed to "Vve. Cesar Vacheron & Cie.". A survey conducted by the two ladies showed that Vacheron undoubtedly belonged to the small group of outstanding manufacturers, above all in the realm of "wearable" small and ornamental watches.

The firm's watches enjoyed particular popularity, especially because of their precision and reliability, which could not be taken for granted from all manufacturers. For that reason it was a wise move to emphasize this precision once again by taking part in the Geneva Observatory's precision contest from their beginning in 1872, in which the firm was able to gain significant awards. But it was also sensible to realize that the respected old name

Left and right: super thin hand-wound movement (height 1.64 mm). The detailed view at left shows the regulator, lever and escape wheel; the hand-setting mechanism is shown in detail at right.

Thin watch movement with automatic winding (height 2.45 mm); the outer part of the rotor consists of 21-karat gold. The detailed view shows the rotor with the attached gold segment as well as the combined stone bearing for mounting the anchor wheel.

Man's gold wristwatch from the "333" line; date indication by the 9; quartz movement. end of wrist.25

of Vacheron & Constantin could be used again. First of all, the two ladies looked for an experienced manager, and in 1875 they hired Philippe-Auguste Weiss as director. After negotiations with Jean-François Constantin, who had been working as an ordinary salesman since 1867, he was now empowered to sign legally binding contracts and documents alone. The business was transferred to the widow Laure Vacheron-Pernessin and Jean-François Constantin and registered as "Vacheron & Constantin". The movement of the workshops, offices and sales rooms, which was done just after the renaming, also resulted from the activities of the two ladies. Since then the Vacheron & Constantin factory has been located on the Quai des Moulins, on the Rhône Island in the heart of Geneva. The "Journal Suisse d'Horologie" stated in a report that the company had attained a leading position in the Swiss watch industry.

The subsequent years of the firm, which now at last was called "Vacheron & Constantin" again, were marked by several decisions that have retained their importance to this day:

– In 1877 the firm's name was changed again, so as to keep the brand name regardless of the ownership of the firm at any time.
– The production of watch movements was completely separated from that of watch cases. Vacheron & Constantin now offered watch movements for sale independent of cases. The customer could first choose a watch case (a particular model); then he could decide which movement he wanted in the case. Through this possibility of making combinations the firm was much better able to grant clients' individual wishes without having to make a much more expensive individual piece.
– In 1880 a new law was passed to protect business samples and especially trade marks. This was made use of immediately by the firm.

Many prominent people were still among the firm's customers. The watches from Geneva had become status symbols that needed to be protected from counterfeits or imitations. Four different trade marks and forms of the firm's name, used on different movements and cases, were registered with the Swiss authorities. All four had something in common: the designation "Vacheron & Constantin, fabricants, Genève", with the added "Horlogerie et boîtes de montres". Two of them showed a particular feature that is still recognized worldwide as the trademark of Vacheron & Constantin, without any further change: the Maltese cross. This "logo" had its origin in the form of the toothed wheel that served in antique watches to regulate the spring tension. Its use as a symbol of the firm's products paid tribute to the creativity and artistic achievements of the first watchmakers. Today the Maltese cross is to be found on the dial. It is also used as a symbol in advertising, on all printed matter and on other parts of a watch (such as on the crown or the attachments).

The subsequent history of the house in brief: in 1882 Catherine-Etiennette Vacheron was a hundred years old. The former

Gold-band watch for men, in yellow and white gold; quartz movement.

This gold man's watch from the "222" line is fitted with automatic winding and date indication.

"gray eminence" of the business was described by the "Journal Suisse" as a symbolic figure of Geneva's fine watchmaking. She died six weeks after her 101st birthday. In 1887 Laure Vacheron-Pernessin, the former director, died. The firm was changed into a stock company and was thereafter called "Vacheron & Constantin, Ancienne Fabrique S. A.". The added "Ancienne Fabrique" disappeared in 1896. Thus the firm finally found the name that has been used to this day. In 1890—fortunately only for a few years—sales problems began, brought on by the makers of Paris fashions. They had decreed that jewelry and watches were no longer wearable. But the Parisiennes did not want to give up jewelry and ornamental watches, and the men of the time also recognized very quickly that a beautiful watch is a piece of jewelry after all. So the watches of Vacheron & Constantin were still ticking at the elegant court of Franz Josef in Vienna as at parties in the palace of Queen Victoria in London in 1899, when Jules Weiss succeeded Philippe-Auguste Weiss as Director.

The year of 1910 was another important date in the annals of Vacheron & Constantin. After several years of experimentation, the factory added the manufacture of wristwatches to its program and involved itself in this new "fashion trend". But the production and development of pocket watches were not neglected. How far the firm had come with these products is shown by a pocket watch made in 1912. It includes an perpetual calendar and indicates sunrise, sunset and equation. During

Man's wristwatch in yellow and white gold, the dial decorated with four diamonds.

World War I Vacheron & Constantin made observation watches and marching compasses, particularly for the US Army. In 1928 an 18-karat gold "Grande Complication" pocket watch appeared, which offered minute repeat, split second chronograph with registers, perpetual calendar, moon-phase indication and alarm.

In 1929 a pocket watch came on the market with the Egyptian royal arms on a gold case. It had a split second chronograph with 30-minute register, minute repetition, perpetual calendar, and was a gift of the Swiss colony in Egypt to His Majesty Fouad I. Finished in 1931 and displayed at the National Exposition in Zürich was a 28-jewel pocket watch (its 12-karat gold case alone weighed 123 grams) with lever tourbillon, winding indication, perpetual calendar, moon-phase indication and split second chronograph. The wristwatches did not come up short among all this development, as shown by a Vacheron & Constantin from 1925, a woman's ornamental watch with many gems on its case and band, equipped with a movement of the 5¾ caliber.

In 1936 a Constantin again assumed the leadership of Vacheron & Constantin. Charles Constantin became president of the firm and called the company the "oldest watch factory that practices interchangeability". He insisted that, just as in Leschot's day, though the metal parts were made by machine, every part had to be worked by hand by a watchmaker.

World War II brought Vacheron and all its competitors great financial problems. The world had other worries. Sales, especially of expensive pieces, declined rapidly. In 1940 Georges Ketterer acquired the majority of the firm's shares and became its president. In the same year the firm put one of the most beautiful wristwatches, with a simple classical design, on the market, a chronograph with 30-minute register, tachometer and telemeter scales and a gold band.

Under Georges Ketterer (until his death in 1987 his son Jacques was the chief) the firm climbed back upward. In the Forties the company, which now concentrated above all on its technical ability and applied it in the area of wristwatch production, built a classically beautiful wristwatch that was reissued in limited numbers in 1985 as the "Jubilee", with an ultra thin mechanical movement. In 1953 the Swiss government gave a Vacheron & Constantin, set with superb dia-

From upper left to lower right:

Man's gold wristwatch with automatic winding, day, date and moon-phase indication.

Man's elliptical wristwatch with automatic winding and date indication; end of the Seventies.

Man's octagonal wristwatch with sweep second and date indication; hand-wound movement; end of the Seventies.

Man's trapeeze-form wristwatch with hand-wound movement; end of the Seventies.

monds, to Queen Elizabeth II as a coronation gift.

After ten years of development, a man's ultra thin wristwatch was offered in the anniversary year of 1955. The movement was only 1.64 mm thick, thinner than an ordinary matchstick. The five-part bridge movement consisting of more than sixty individual parts also had a diameter of only 20.8 mm and offered a whole series of other special features. The parts of the movement were so perfectly refined by hand by the watchmakers that they proved to be fully insensitive to corrosion. In addition, the watch was constructed so that shock resistance could be omitted. The running precision was so perfect that this watch did not need to be readjusted after a cleaning or lubrication. On the other hand, the original regulation took a master watchmaker a whole month.

The escape wheel had a thickness of only 0.14 mm (outside and inside), and 0.09 mm in between. The ticking of such a watch sounds especially high-pitched, almost crystalline, and very soft, because the consumption of power, thanks to the construction, is very slight.

In the jubilee year of 1955 there was also a meeting of the world's most important statesmen in Geneva's "Palace of Nations" on July 18 to confer on world peace: Nikolai Bulganin (Soviet Union), Prime Minister Anthony Eden (Great Britain), the American President Eisenhower and the French chief of government, Edgar Faure. When the conference ended, twenty citizens of Geneva gave each of these statesmen a Vacheron & Constantin with an inscription: "May this watch always show

Man's octagonal wristwatch with central band attachments; quartz movement.

Man's octagonal wristwatch; stepped case rim and band attachments.

happy hours—for you yourself, your people and the peace of this world."

In 1967 Vacheron & Constantin took part in the "competition for the thinnest watch", producing a new best achievement in the form of an automatic movement with a height of only 2.45 mm! It took six months' work to build such a movement. For the sake of weight, the rotor circle was made of 21-karat gold. The mounting of the rotor, which was not—as usual—in the center of the movement, but outside it, consisted of a circular beryllium-bronze rail that rested on ruby rollers. This gave two advantages, namely decreased height and reduced friction.

Vacheron & Constantin sees itself today as the "oldest watch factory in the world that has built watches without interruption". That the ownership conditions have changed over the years is neither confirmed nor denied by Vacheron & Constantin. Newspaper announcements in which a former oil minister and an Arab sheik have had their say are not mentioned.

The development of quartz watches has also found a place in the factory, for quartz technology definitely has its advantages in the manufacture of watches. Ornamental watches, for example, that are equipped with quartz movements will still show the right time after not having been worn for some time. The percentage of quartz watches in the total production corresponds to that of women's watches: "about 35%", they say at Vacheron & Constantin. They would be happy to see the demand for quartz watches decrease and the traditional mechanical masterpieces sell in greater numbers. For two years they have not sold any electronic watches to Japan, the source of the quartz invasion, they say, not without pride. And Japan is an important market. Europe and the Far East are balanced in importance as sales markets. Italy leads Europe, ahead of France, Great Britain and Switzerland. The leader in the Far East is Japan, ahead of Hong Kong and Singapore. The Unites States holds third place. Switzer-

Man's octagonal wristwatch with stepped gold case; diamonds mark the hours on the dial; automatic movement.

Man's round gold wristwatch.

land has an additional international significance as a sales market: on account of the strict customs restrictions of the South American states, most sales to South American customers take place in Geneva, for these clients can buy their watches much more reasonably in Switzerland than in South America. And in Switzerland—here too, particularly in Geneva—the most Arabian sales take place. Vacheron & Constantin does not need to go to the sheiks, for the oil billionaires come to the Rhône Island.

In France the firm has its own branch, that also makes Cartier watches out of Vacheron & Constantin movements. For some years the German market has also been served by a branch, although it admittedly does not yet play any great role in sales figures. But that is due to change soon. There is a good rate of growth, they assure us in Geneva, and the German market is interesting. For that reason the firm's activities are being increased bit by bit, because the Federal Republic is now the greatest potential market. But the firm is not using aggressive advertising, preferring truly exclusive events such as an international tennis tournament in Berlin with a watch as the prize.

Some 450 agents worldwide retail the watches of the house. That is not many more than at the beginning of this century, when there were 427. According to the prevailing estimates, it is not likely that the number of retailers will climb. Only jewelers who have the class and the financial means, and who fit in with the style of the house, can become agents. This limits the numbers of potential candidates to a great degree. With a year's production of five to six thousand watches, which is not going to be increased, high quality and precision can be guaranteed in the future too. Sixty employees work in production (fifty of them are watchmakers), and twenty more in warehousing, sales and administration, even though not all the raw movements are built by the firm itself any more, but for the most part bought from LeCoultre. But they are traditionally reworked and refined, piece by piece.

Nothing is said about prominent people who own a Vacheron & Constantin. Only this much has leaked out: Queen Elizabeth II owns one, and so do a list of other personalities whose names are immortalized in the firm's Golden Book in Geneva: Prince Edward of Britain, the Sultan of Morocco, actor Pierre Fresnay, actress Merle Oberon and author Andre Roussin are among Vacheron & Constantin's customers. Princess Diana wears

From the "Struktura" line comes this mechanical gold-band watch with a chased movement, the balance of which is visible at the front.

This detailed view shows a chased skeleton movement.

Man's wristwatch in yellow and white gold.

Woman's wristwatch from the "222" line, decorated with diamonds; quartz movement.

From the cooperation between Vacheron & Constantin and the painter Raymond Morette comes the "Kallista" ("most wonderful"), a unique wristwatch that is also the most expensive in the world.

a "Lady Kalla", which was ordered by the legation of the United Arab Emirates in Geneva on the occasion of her marriage to Prince Charles. Today this is the most expensive watch made by the firm which, as has been said, stresses the fact that its products are not a question of price.

That is why the sporting "333" line exists in women's and men's versions, in steel, steel-gold and 18-karat gold, with automatic or quartz movements. The wristwatches with perpetual calendars are real masterpieces, and have mechanical movements without exception. They come in gold or platinum cases, but also as skeleton watches. The other mechanical skeleton watches with hand or automatic winding all have gold cases (but can be had with leather bands). "Half"-skeleton watches form the "Struktura" line. Gold cases can be taken for granted here too. The second sporting line, the "222", is made to resemble a screwed-in porthole of a ship; here steel is also used. The models are naturally watertight, thanks to a "double crown". Selling for about 100,000 marks is the man's "Harmony" watch, set with numerous diamonds, one of the most expensive quartz watches of our time.

The women's watches with baguette or "barrel" calibers are the real wonders. Here the main goal was to make the movement so small that it can be hidden completely under the jewelry. Thus the baguette movement measures only 17.6 by 6.3 mm, with a height of 3,5 mm. The "barrel" movement measures 15.4 by 13 mm with a height of 2.6 mm. These models, as well as the other watches made by this firm, some of which have been built for decades in unchanged classic form, are protected by perfect service, just as the firm's earlier products were. Part of the guarantee assures that it will not be the dealer, but the manufacturer who is responsible for everything.

Drawings for every watch made since 1840 still exist, with case and movement numbers, dates of manufacture and sale, and a note of the retailer who sold the watch. Vacheron & Constantin also have a large supply of spare parts, enough to maintain every watch for at least twenty years. And if the watch is older, replacement parts will be made new. In any case the customer receives a cost estimate, as such individual work naturally can become expensive. At any time one can get information about a watch if one sends in the case and movement numbers. And the specialists at Vacheron & Constantin also give—naturally without obligation—information about the present-day value of an antique watch.

It is not impossible nowadays that the firm might even make an offer to buy a watch, in order to add to its own private collection of several hundred very rare pieces (some from the Sixteenth Century) and establish a small company museum.

In October of every year, the firm's proprietors, directors, and the nine most important dealers meet and look into the future. From the suggestions made by designers and master watchmakers, they decide what shall be produced in the fall for the coming year. In limited numbers, of course, so that the art of perfection in all its exclusivity will live on at a rate of 38 to, at the busiest times, no more than 43 watches per day.

Extra-thin and watertight wristwatch with quartz movement, from the "Harmony" line. The model shown is made of yellow gold and steel.

From the line "Les Absolues de Vacheron & Constantin" comes the "Lady Kalla" wristwatch for women.

Man's rectangular gold wristwatch from the "Harmony" line, with quartz movement.

TECHNICAL TERMS

The notations (») indicate terms explained elsewhere in this list.

Analog Indication
  Indication by hands. Analog means "corresponding" or "in the sense". The hands stand according to the time. The term "analog watch" was taken from the field of electronics with the spread of the quartz watch.*

Antimagnetic Watches
  These watches are extensively protected against magnetism. Magnetic fields, which can be created by household appliances, electric motors, etc., do not disturb these watches. But very strong magnets can bring even these watches to a stop. Quartz watches are not disturbed by magnets. But very strong magnets (magnetic keys) can briefly impede the running (the step-switch motor).*

Astrolabe
  Also called planetarium, analemma or angle knife. Instrument for measuring angles in degrees, minutes and seconds of arc. The invention of the astrolabe for orientation at sea goes back to, among others, Martin Behain of Nürnberg (circa 1459-1506).

Automatic Winding, Wristwatch With
  Movements of the hand move a swinging weight (rotor or pendulum). An apparatus makes the winding drive always turn in the same direction to wind the mainspring. A sliding coupling (drag spring) prevents overwinding the mainspring. Automatic watches usually are more precise than watches with hand winding, because they usually run at full spring strength.*

Baguette Movement
  Watch movement in a longish rectangular form, its length at least three times its width. Baguette movements were in style for wristwatches chiefly in the Twenties and Thirties.

Balance
  The running regulator of the machanical wristwatch and pocket watch. The balance wheel has the role of a swinging weight, the hairspring must always bring the balance back to its resting position. The swinging weight and the returning power of the hairspring are always set against each other, so that the desired number of swings is attained. The classic frequency is five beats per second. To improve accuracy, the balances of modern wristwatches beat faster (up to ten beats per second). The balance is mounted in jewel (ruby) bearings. In every bearing a (») hole jewel and a (») cap jewel are used. The pivots of the balance arbor are very thin (about 0.1 mm) to minimize friction.

Batteries for Wristwatches
  The smallest dry cells, also called button cells. They should be air- and watertight and, despite their smallness (diameter about 6-10 mm, height 3-5 mm), should power the watch at least a year. Modern quartz watches use silver oxide cells (the silver oxide is the plus pole and binds the hydrogen originating in the cell; the minus pole is made of zinc). The voltage of these cells is about 1.5 volts and is very constant. Lithium cells last longer and will probably be probably be used in the future as power sources for wristwatches.*

Beats: (») Balance

Beryllium Balance
  Balances of modern watches are made of a beryllium-based alloy; thus this term is only partly correct. Such balances have great hardness and firmness, have a golden yellow shine and scarcely oxidize.*

Bezel
  Meanings vary; in watchmaking the bezel is the ring that holds the crystal. The watch crystal is pressed into it, the bezel itself then sprung against the middle of the watch case.

Bimetallic Balance
  These two-metal balances are meant to prevent influences of temperature variation from affecting the running of the watch.

The balance ring is made of brass outside and steel inside. The ring is also cut in the vicinity of the shank. With rising temperature the brass expands faster than the steel, the ring ends bend inward, the inertia of the ring decreases. Since the hairspring also becomes weaker with rising temperature, the running of the watch remains the same. Today the temperature error is prevented by hairsprings that do not weaken (Nivarox).*

Block
Carrier of an upper wheel bearing. Blocks are screwed to the (») plate and held exactly in position, like (») bridges, by setting bars.*

Breguet Hairspring
It prevents the hairspring from swinging farther to one side (eccentric swinging). The last loop of these springs is raised up and bent into a precisely defined curve. Breguet hairsprings were used in precision pocket watches and can be found in good wristwatches. (A. L. Breguet, Paris, 1747-1823).*

Bridge
It carries the upper bearings of two or more wheels. Bridges are screwed to the plate and their position is secured by setting bars.*

Cabinotier
Seventeenth and Eighteenth Century term, as master watchmakers still built complete watches with the tools in their "cabinets".

Calendar
See (») perpetual calendar, Gregorian or four-year calendar.

Caliber
Term for the type of movement. Caliber designation consists of manufacturer, reference number andor size.*

Cannon Pinion
On the minute arbor is the turnable cannon pinion which carries the minute hand. The canyon pinion drives the minute wheel. The hour wheel pipe is mounted over the cannon pinion and driven by the minute wheel pinion. The hour hand is mounted on the hour pipe.*

Cap Jewel
It reduces the friction of the bearing, limits arbor play. Cap jewels are always used in balance bearings, sometimes in wheel train and other bearings (second wheel, third wheel).*

Center-of-Gravity Error
The balance is not balanced. A very small imbalance disturbs the running of the watch. (The watch runs differently in different positions.) The balance and its arbor are mounted on two rubies.*

Chasing or Chiseling
Cutting of ornaments or figures out of originally smooth surfaces for decoration.

Chronograph
A normal watch with additional apparatus for measuring (stopping) short times. Mechanical chronographs can also have minute or hour registers as well as a "split second" mechanism for intermediate stopping. Electronic digital watches are often equipped as chronographs.*

Chronometer
Watches with very high precision, which fulfill the requirements of official testing agencies. Testing is usually done over a period of sixteen days. The watches are tested for, among others, position and temperature errors. The term "chronometer" is copyrighted for mechanical watches. There is also testing of quartz watches.*

Cleaning
Whenever a watch is overhauled, the old, used, usually hardened oil must be removed, since even small traces of old oil can cause disturbances. The watch is dismantled and rinsed several times with liquid fat-solvents in a cleaning machine.*

Compensated Balance
Meant to counteract the influence of temperature changes on the running of a watch; (») bimetallic balance.*

Crystals for Watches
Plastic is often used for watch crystals, as it does not break and is easily pressed into the crystal ring. The crystal fits very well into the sharp rim of the ring to give a very tight seat. In watertight watches a metal ring is set on the plastic crystal to make a tighter seat. Plastic crystals scratch very easily. Crystals of hard (mineral) glass are more resistant but have to be thickened. Sapphire glass is scratchproof. Wristwatches made before 1940 often have glass crystals, which break very easily.*

Cylinder Watch
A small hollow cylinder is used as its escapement; about half its effective part is removed. In wristwatches this cylinder is about 1 mm in diameter; its walls are about 0.1 mm thick. Since the cylinder directly bears the weight of the balance, it is very sensitive to shock. Wristwatches with cylinder escapement were still built after World War II. Repair of cylinder watches is rarely possible today, as spare parts (cylinders) are rarely found.*

Dayglow Colors
Self-glowing paints contain traces of radio-active substances which make the phosphorescent basic material (zinc sulfide) glow. Alpha radiators were formerly used, now a beta radiator (tritium) is used. Dayglow colors are sparingly used in modern watches because of the danger of radiation.*

Deviation
Daily deviation is the discrepancy from the real time within 24 hours. This discrepancy is stated in seconds per day. In very good mechanical watches the deviation is only a few seconds per day. Quartz watches are more precise, deviating only a few seconds per month.

Digital Indication
Indication by numerals. Mechanical wristwatches with digital indication were known in the Twenties, and became common through the introduction of microelectronics in time measurement (quartz watches) in the Seventies and Eighties.*

Display
Electro-optic indication. In digital watches only (») Liquid Crystal Display is still used. There are also displays which simulate hands (electronic analog watch).*

Diver's Watch: (») Watertight Watch

Ebauche
Term meaning the raw movement, the heart of a watch. Unlike (») manufacturers, many (») finishers buy raw movements from ebauche producers. The components of a raw movement (plates, (») bridges, (») blocks, (») wheel trains, (») hands, etc.) can be purchased in various grades of preparation, for example, with or without jewels. Because of the work and facilities involved, raw movements are made by only a few specialized producers. Other components (balances, anchor parts, hairsprings, jewels, mainsprings, screws) are also usually made by specialized firms.

Electric Wristwatch
Only electric components such as contacts, coils, condensers or resistors are used in electric watches, including wristwatches, in which the balance closes an electric circuit over a contact for a small magnetic pulse that gives the balance an impulse. The first electric watch was marketed by Lip (France), Hamilton and Elgin (USA) in 1952. The principle of such watches was long known. The technological problems in producing electric wristwatches were in the small impulse which needs a very thin wire, and in the contact, for which only very little power is available. The making of usable batteries also caused problems. The Lip, Hamilton and Elgin watches were soon followed into production by those of

Epperlein (Germany) and Landeron (Switzerland, Ebauches Konzern).*

Electro-erosion
A process used since the Fifties to work hard or relatively inflexible materials. In electro-erosion the properties of the electric spark are used to remove material from the piece being worked on.

Electronic Wristwatches
Watches that have semiconductor elements like transistors or integrated switching. Quartz watches are electronic watches.*

Equation
Time comparison, defined as the difference between real and mean sun time. The real sun time is determined by the revolution of the earth around the sun. Because of its elliptical orbit and the tilt of the earth's axis, the length of the real sun day varies. Mean time is an average time, from which the time equation is derived in comparison with real time. Equation varies between extremes of –14.3 (February 11) and +16.4 minutes (November 3). Four times a year, on April 16, June 14, September 1 and December 25, real and mean sun times are equal.

Escapement
It consists of escape wheel and lever. The escapement must regulate the quick, irregular running of the movement, keep the wheel train turning in time with the balance, and transmit power to the balance.*

Finisher
Unlike the (») Ebauche producer, the "Etablisseur" uses only purchased parts (raw movements, escapement and swinging systems, bearing jewels) to make finished movements (» termination), sets them in cases and markets the finished watches.

Formed Movement
Watch movements that are not round; usually rectangular, long thin (» Baguette) or oval movements.*

Four-Year Calendar
Unlike the true perpetual calendar, a four year calendar does not give the correct date from one leap year to the next. On February 29 of every leap year it must be manually corrected.

Gears
They are fraised for watches (in a process of rolling). Arc toothing, derived from cycloid toothing, is used. Cycloids are rolling arcs that are formed on a firm circle by rolling. Gears with from 20 to 100 teeth are made of brass. Wheels with few teeth (6-12) are called pinions; they are made of steel and hardened.*

Glucydur
Alloy for balances and mainsprings, a copper alloy with 2 to 3% beryllium added. (») Beryllium Balance.

Gregorian Calendar
In 1582 Pope Gregory XIII, after years of study, introduced a calendar reform setting the length of the year at 365.2425 mean sun days. The introduction of the Gregorian calendar largely eliminated astronomical irregularities of former time systems, and the calendar matched the actual movements of the heavens. Along with the varying month lengths, use of February as the leap-year month and four-year leap-year rhythm of the Julian calendar, Pope Gregory XIII decreed that only those full century years should be leap years that are divisible by 400. For this reason, 2100, 2200, 2300, 2500, 2600, 2700 etc. will not be leap years. But even this system does not eliminate all errors, which will add up to one day in the course of 3300 years.

Hole Jewel (Bearing Jewel)
The bearing jewels are made of synthetic ruby and drilled with fast-turning copper or bronze tools bearing diamond cutters. The holes for the bearing pivots are bored mechanically or by laser beams. Today hole jewels are pressed in place; they used to be set. Tolerances in making hole jewels are

very small. Deviations may amount to only 0.0025 mm.*

Isochronism
In isochronic swinging, the swing duration is independent of the swing distance and the swings are all equal. Balances are to swing isochronically. This requirement is basically achieved today.*

Jewels
To decrease friction in the most important bearings, pallets and rollers of precision watches, jewels are used. In the past, natural jewels (rubies or sapphires) were used; modern watches use synthetic jewels. There are bearing jewels (» hole jewels), (») cap jewels, pallet and roller jewels. The number of jewels used in a watch does not always indicate special quality or value. A hand-wound precision watch usually has at least 15 functioning jewels: ten hole jewels, two cap jewels, two pallet jewels and one roller jewel. More complex watches, such as those with automatic winding, chronograph or repetition striking, have correspondingly high numbers of jewels.

Lever Escapement
Also called club-tooth, Swiss anchor or free anchor escapement. The anchor bears two stones (pallets), the third stone, the roller jewel, is set in the balance and works with the fork of the anchor. This escapement is used in good mechanical wristwatches and goes back to Thomas Mudge (1715-1794).*

Light Emitting Diode (LED)
When tension is applied, this diode lights up. LED are used for digital indication, measuring instruments, calculators etc., but are not used in modern digital watches because of high power consumption. To read the time from LED wristwatches, the indication has to be turned on by pushing a button. Long-term indication is impossible because of the high power consumption.*

Liquid Crystal Display (LCD)
This form of display is often used in quartz watches because it uses very little power. A battery can power the watch for several years. If the watch has a bulb, this uses the most power. Liquid crystals show a crystalline structure despite the condition of the liquid (molecules are in order). If electric tension is applied, the liquid crystal changes its optical qualities. Dark symbols are visible on a light background.*

Ligne
Old measure of length: 1 ligne (1''') approximately 2.256 mm. Lignes are still used to give the sizes of movements. The most commonly used sizes for wristwatches are between 5.5 and 13 lignes.*

Lubrication
Fats and oils are used to lubricate watches, oils for the wheels, fats or pressure-resistant oils for the springs. Watch oil must stay in its place and may not harden if it is to remain fully effective. Only very small drops of oil are used in bearings.*

Manufacturer
A manufacturer is a watch producer who is both (») ebauche maker and (») finisher, with both raw movement production and finishing taking place in the same company. The movements are called manufacturers' calibers.

Marine Chronometer
Highly precise type of watch, also called ship's chronometer, wherein the special escape wheel is detached from the balance by a sprung detent. Mechanical chronometers were used before the universal introduction and use of time transmissions and later quartz watches for correct clock time on ships, needed to determine their positions.

Mechanism
The apparatus added to complex watches, such as for repeat, or perpetual calendar.

Megaquartz
The quartz of this watch swings at 4.19 megahertz (4,194,304 swings per second).

This high frequency gives even greater precision. Only megaquartz movements are now used in large clocks. Power supply is still a problem in wristwatches, since the high frequency needs many dividing steps and thus more power is consumed.

Module
A separate component functioning separately. In electronic watches the bottom plate with all its electronic components is called a module. In toothed wheels, a module means an additional size that makes wheel calculation easier.*

Moon-Age Indication
Indication showing the number of days since the last new moon. In a synodic month, one lunation, the time from one new moon to the next, equals exactly 29 days, 12 hours and 44 minutes. The indication is usually done with a 59-tooth wheel that turns once in two lunations and whose position can be read through a window in the dial, which is completed by a scale graduated to 29.5 days. Depending on their construction, watches with moon-age indication have a yearly error of one minute to about eight hours.

Moon Phases
The moon goes through its phases: new moon, first quarter, full moon, last quarter, and back to another new moon, caused by the location of sun, earth and moon, in one lunation of about 29.5 days (» Moon Age).

Nivarox
Special alloy for hairsprings. Nivarox hairsprings are not influenced by temperature changes. Nivarox is an iron-nickel alloy with added chromium, beryllium and other metals. Nivarox is as hard and elastic as steel, non-rusting and non-magnetic. The Nivarox hairspring makes the old compensated balance superfluous.

Perpetual Calendar
Calendar which automatically includes the different month lengths in normal and leap years. Wristwatches with perpetual calendar usually indicate date, day and month. Some models also have leap-year indication. Perpetuality applies to the nature of the Gregorian calendar, wherein the formula lasts only until February 28, 2100. Since 2100, as opposed to the usual four-year sequence, is not a leap year, the calendar must be switched manually on March 1, 2100. Likewise, 2200, 2300, 2500, 2600, 2700 etc. will not be leap years. The perpetual calendar requires a complex mechanism with numerous wheels, levels and ratchets, which is attached to a movement.

Piezoelectric Effect
Certain crystals like quartz or tourmaline show this effect. If magnetism affects them, electric charges arise on their surfaces. Electric tension can briefly reform the crystal. Electric oscillations make the crystal oscillate and again create alternating tension on the surface. The regulator of quartz watches, a quartz plate (fork), oscillates as a result of this effect.*

Pillar Movement
The wheel train runs between two plates; the distance between them is fixed by pillars. Pillar movements are used in cheap watches. In better watches only one plate is used, to which (») blocks and (») bridges are screwed to hold the bearings.*

Pin Lever
The anchor of a lever escapement bears small ruby jewels or pallets. Instead of pallets, steel bars contact the brass escape wheel with the pin lever. Pin lever escapements are used in simple watches (» Roskopf Watch).*

Integrated Circuit (IC)
Electronically active layers are applied to a small crystal plate of silicon or germanium to make transistors, diodes, resistors or condensors. These elements are connected by leaders. It is possible today to fit up to 100,000 such elements in one cubic centimeter. These integrated elements made quartz watches possible.*

Plaqué
Electroplated material, used for watch cases.*

Plate
The movement plate of a watch holds bearings for the wheels, while the second bearings are in (») blocks or (») bridges.

Quarter Second Stop (Seconde Foudroyante)
Auxiliary second hand that turns around its axis in four or five springs per second.

Quartz
Mineral, chemically SiO2, hardness 7, known in purest form as rock crystal. Quartz shows the (») piezoelectric effect, because of which quartz can be made to oscillate by electronic switching, keeping its frequency very constant. Oscillators are made today of synthetic quartz crystals; the quartzes used in the most modern watches have the form of a tiny tuning fork (only a few mm long).*Raw Movement: see Ebauche.

Regulation (Fine Setting)
Regulating a watch consists of observing its daily (») deviation in various positions and temperatures and adjusting them accordingly. Depending on the quality and desired accuracy of a watch, varying regulating procedures are used. The usual regulation of a good watch consists of testing in dial-up (lying) and crown-up (hanging) positions. The deviations between these positions are usually 30 seconds a day at most. In officially prescribed precision regulation, watches are tested and adjusted in at least five positions and at two different temperatures. A requirement of good regulation is an exactly balanced (») balance, since (») center-of-gravity error would otherwise occur. In most cases, correction of mechanical watches is done by using the regulator, which changes the effective length of the hairspring. The art of regulating mechanical watches consists in principle of keeping the number of swings of the balance or hairspring as constant as possible despite disturbance from external influences such as temperature and position changes. When the frequency changes, errors result. Quartz watches are regulated by using a trimmer.

Regulator
Used to regulate watches with balances. The last part of the hairspring is led between two rods or angles. By turning the regulator, the effective length of the hairspring, and thus the swing duration of the clock, is changed.* (» regulation).

Remounting
Building the raw movement into a watch ready for use (with dial, hands, case and balance). Formerly, when series production was being developed, movements were assembled provisionally, then dismantled, production errors corrected, and the movements reassembled (remounted). The factory watchmaker is thus called a remounter.*

Repassage
The last testing of a watch before delivery.

Repeating or Repeat Striking
Watch complexity that lets the actual time be indicated more or less precisely by striking on coiled gongs. According to how the striking mechanism is built, there can be quarter repeating, half-quarter repeating (7.5 minutes), 5-minute or minute repeating, of which minute repeating is the most complex and exact form. Pocket watches with repeating were quite widespread, but wristwatches with this complication are rare.

Rolled Gold
A thin (a few hundredths of a mm) sheet of gold is pressed on the material (Tombak), glowing hot. The desired thickness is attained by rolling. Rolled gold (double) is used for watch cases.*

Roskopf Watch
This term refers to simple wristwatches, usually without jewels. A pin lever escapement, also without jewels, is used. The (»)

wheel train is not arranged in the usual order. The hands are not on the (») pipes in the center of the movement, but on a bar in the plate. Such a movement is also called a flying-hand movement. (G. F. Roskopf, 1813-1889).*

Rotor
The rotating weight of an automatic wristwatch which winds the mainspring via a toothed-wheel drive. In quartz watches, the rotating permanent magnet of the step-switch motor.*

Ruby
Material for bearing jewels; today synthetic ruby is used. The purest oxide is melted in a gas flame. Added chromium oxide gives the red color.*

Sapphire
Formerly used like ruby as a bearing jewel. Scratchproof watch glasses are made of synthetic sapphire.*

Scratchproof Watch
The case of a watch is made of hard metal (an alloy similar to that used for steel- and stoneworking tools). The glass of the watch is synthetic sapphire (hardness 9).

Screws
The smallest screws are used in wristwatches. The smallest thread used had a diameter of only 0.3 mm. 10,000 of these screws weigh one gram.*

Self-Compensating Hairspring
Balance springs made of spring steel tend to change their elasticity with changing temperature, changing the precision of a watch. Until the 1930's, precision watches were made with (») bimetallic balances to counteract this physical tendency; their rings were made of two different metals, usually steel and brass. These rings, cut at the shanks, considerably equalized the temperature error of the hairspring. Much research led to the introduction in the Thirties of a hairspring that was made of an alloy and thus able to equalize temperature variations itself. These self-compensating hairsprings went on the market in 1933 under the (») Nivarox name, and were soon widely used because of their excellent qualities. The bimetallic compensated balance, which was hard to manufacture, thus became superfluous.

Shock Resistance
Protects the sensitive balance arbor and bearing of a watch from breaking under shock. Generally the bearing jewels are sprung so that the jewel and balance arbor are not damaged. The terms "shock resistant" and "shockproof" are copyrighted. A watch so designated will not be damaged or show any great deviation by falling onto an oak surface from a height of one meter.*

Skeleton Watch
Plate, blocks and bridges (and barrel) are cut out, leaving only thin bars. The wheels and functioning of the watch are visible. Making skeleton watches calls for the greatest handworking ability. Watches of this kind are thus very valuable.*

Split Second Chronograph
Chronograph with additional hand (sweep hand), independent of the actual chronograph hand, that can be stopped, for example, to register time-outs at sporting events. At the end of the intermediate stop, the sweep hand springs back into agreement with the chronograph hand. This process is repeatable as often as wanted. Wristwatches with split second chronographs are very rare because of the additional construction and resulting high price.

Spring Winding
Mainsprings were already used circa 1500 as power sources of clocks. The mainspring gives its fullest power (turning moment, which then decreases more or less regularly) when fully wound. Modern mainsprings for wristwatches are about 0.1 mm thick and 200 to 500 mm long. Certain alloys are

very unlikely to break (unbreakable springs).*

Stave Movement
The wheels of the watch are mounted between thin plates in the form of rods so that the working of the watch is visible.

Sweep Hand: see Chronograph-Rattrapante.

Termination
Process of finishing a watch movement, consisting of testing all parts, doing necessary fine work and completing a functioning watch.

Thermal Watch (Thermatron)
This watch uses a thermal element as its power source, consisting of a soldered joint linking two metal wires. When the wires reach different temperatures, electric tension results, growing with increasing temperature difference. Body temperature is the source of the higher heat; the upper surface of the case provides the lower heat. The thermal element charges a small battery.*

Timing Machine
Regulating apparatus for watches. The sound of the watch is picked up by a microphone and compared to the impulses of a quartz watch. The scale shows the deviation of the watch in seconds per day.*

Tourbillon (Turning Frame)
Construction invented by Abraham-Louis Breguet in 1795, patented 1801, in which the whole escapement system is housed in a turning frame to equalize the timing errors that occur in different positions, especially in a vertical position. According to the layout of the wheel train, the tourbillon rotates once in a given time (usually 1 minute).

Tuning Fork
Oscillates 360 times per second and regulates running. It carries two small magnets and is kept oscillating by transistor switching. On a soft spring it bears a ruby that moves a switching wheel with a diameter of only 2.4 mm. Its circumference has 300 teeth, each 0.01 mm high and 0.03 mm wide. Switching can only be adjusted under a microscope. The tuning fork's high frequency gives very great precision. Tuning-fork watches provided a new way of measuring time, of highest quality. The fast development of quartz watches (the higher regulator frequency gave greater precision) prevented more widespread use of these watches. Tuning-fork watches with visible movements were very popular.*

Waterproof Watches
There are no absolutely waterproof watches. Watches called "waterproof" must be able to withstand the water pressure of a one-meter depth for one hour. A series of quick testing methods were developed for the watchmaking workshop.—Divers' watches must withstand much higher demands (pressure).*

Wheel Train
A normal wristwatch has five gear-pinion pairs. The barrel drives the minute pinion, the minute gear drives the intermediate pinion, the intermediate gear drives the second pinion, and the second gear gives its power to the escape wheel pinion.*

Terms marked * were taken from the works of Engineer R. Proidl and Anton Kreuzer, Carinthia Publishers, with their kind permission.

BIBLIOGRAPHY

Alfred Auguste Ernaut (1817-1889):
Tempus ex machina.

Audemars Piguet: Kunstwerke unserer Zeit, 1986.

Baume & Mercier: Wie man Baume & Mercier verkauft, Geneva.

Berner, G. A.:
Dictionnaire Professionel Illustre de l'Horlogerie, La Chaux-de-Fonds, 1961.

Breguet: Breguet heute, Paris, 1986.

Breguet:
Ivanovich Kuprin (1870-1938).

Brunner, G. L.:
Armbanduhren mit "ewigem Kalender"—ein Vergleich, in Alte Uhren, Vol. 41985, pp. 41-61.

Brunner, G. L.:
Armbanduhren mit Repetitionsschlagwerk, in Uhren, Vol. 41986, pp. 65-79 & Vol. 31986, pp. 50-58.

Brunner, G. L.:
Audemars Piguet—Manufacture d'Horlogerie, in Uhren, Vol. 41986, pp. 9-40.

Brunner, G. L.:
Blancpain—Uhrmacherei mit 250jähriger Tradition, in Uhren, Vol. 11988, pp. 9-28.

Brunner, G. L.:
Die Armbanduhr mit Minutenrepetition aus dem Hause Blancpain, Lausanne, 1988.

Bruton, E.:
Uhren—Geschichte, Schönheit und Technik, Eltville, 1982.

Callwey Verlag: Alte Uhren, München.

Carrera, R.:
Symphonie für eine Equation, La Chaux-de-Fonds, 1985.

Carrera, R.:
Tourbillon with three Golden Bridges, Girard-Perregaux S.A., La Chaux-des Fonds, 1983.

Demeter, U.:
Eine Mumie wird zum Weltstar, Jardin des Modes, 1986.

Fahrni, O.:
Verschlossen wie eine Oyster, Bilanz, 1986.

Gerald Genta: Time to Love, Geneva.

Geschichte eines Namens
Corum, Ries, Bannwart & Co., La Chaux-de-Fonds, 1982.

Hassenkamp, S.:
Die Klunker-Klan, Stern, Hamburg, 1984.

Helmut Teriet GmbH: Düsseldorf, 1986.

HL:
Hors Ligne Publishing S. A., Geneva, 1985.

HL Special: Swiss Horology,
Hors Ligne Publishing S. A., Geneva, 1986.

Huber, M. / Banbery, A.:
Patek Philippe Genève, Zürich, 1982.

Huber, M. / Banbery, A. / Brunner, G. L.:
Patek Philippe—die Armbanduhren, Geneva, 1988.

Jagger, C.:
Wunderwerk Uhr, Zollikon, 1977.

Kahlert, H. / Mühe, R. / Brunner, G. L.:
Armbanduhren—100 Jahre Entwicklungsgeschichte, 3rd edition, München, 1986.

Kreuzer, A.:
Die Armbanduhr, Klagenfurt, 1983.

Kreuzer, A.:
Die Uhr am Handgelenk, Klagenfurt, 1982.

Kreuzer, A.:
Faszinierende Welt der alten Armbanduhren, Klagenfurt, 1985.

Kurzweil, A.:
Man of many faces, Washington DC, 1985.

Meis, R.:
IWC-Uhren, Klegenfurt, 1985.

Nadelhoffer, H.:
Cartier—König der Juweliere, Juwelier der Könige, Herrsching, 1984.

Orbitex AG: Bild der Schweiz, Zürich.

Piaget:
Piaget—ein Name, eine Familie, ein Stil, S. A. Ancienne Fabrique Georges Piaget, Geneva.

Proidl, R.:
Glossar technischer Begriffe, in Kreuzer, A.: Die Uhr am Handgelenk, Klagenfurt, 1982.

Räther, H.:
Münchener Merkur, München, 1986.

Rolex:
Collection Hans Wilsdorf, Montres Rolex S. A., Geneva.

Rolex:
Das Leben Benvenuto Cellinis, Geneva, 1980.

Rolex:
Hundertjahrfeier der Fabrik 1878-1978, Manufacture des Montres Rolex S. A., Biel, 1978.

Rolex:
Von Stufe zu Stufe, Rolex-Vademecum I, Geneva.

Rolex:
Die Entwicklung der Armband-Chronometrie, Rolex-Vademecum II, Geneva.

Rolex:
Wie die wasserdichte Uhr entstand, Rolex-Vademecum III, Geneva.

Rolex:
Die Geschichte der automatischen Uhr, Rolex-Vademecum IV, Geneva.

Saturday Review, 1985.

Schmelzer, B.:
Wie alt ist meine Taschenoder Armbanduhr?, Düsseldorf, 1986.

Schulte, C.:
Lexikon der Uhrmacherkunst, Bautzen, 1902.

Singapore Business:
A Man to Watch, Anna Teo, 1986.

Stolberg, L.:
Lexikon der Taschenuhr, Klagenfurt, 1983.

Tölke, H. F.King, J.:
International Watch & Co., Schaffhausen, Zürich, 1986.

Uhren Huber:
Die Cartier-Uhren, München, 1982.

CREDITS

Uhren Huber:
Uhren Huber, München, 1985.

Uhren, Juwelen, Schmuck, 1986.

Vacheron Constantin:
Vacheron Constantin, Geneva, 1973.

Welt am Sonntag: Hamburg, 1986.

Wilsdorf, H.:
Rolex, Geneva.

Archive photos of the manufacturers:

Audemars Piguet, Baume & Mercier, Blancpain, Breguet, Cartier, Chopard, Corum, Ebel, Gerald Genta, Girard-Perregaux, IWC, Jeager LeCoultre, Patek Philippe, Piaget, Rolex, Ulysse Nardin, Vacheron & Constantin.

Armbanduhren, Callwey-Verlag, München.

Auktionshaus Dr. H. Crott, Aachen.

Auktionshaus Habsburg, Feldmann, Geneva.

Brunner, Gisbert L., München.

IWC-Uhren, Karinthia-Verlag, Klagenfurt.

Markwalder, Ch., Basel.

Mercier, C., Geneva.

Patek Philippe—die Armbanduhren, Antiquorum, Geneva.

Swiss Horology, Geneva.

Teriet, H., Düsseldorf.

Trielus, E., München.

Uhren Huber, München.

Voithenberg von, G. u. E., München.

The brand names mentioned in this book are registered trade marks of the manufacturers.

The largest collection of vintage wristwatches and pocketwatches in America.

Collector's catalog: $10/annual subscription (Four issues per year)

AARON FABER GALLERY

666 FIFTH AVENUE (ENT. ON 53RD ST.) NEW YORK, NY 10019 212-586-8411

YOUR WEST COAST SOURCE FOR FINE EUROPEAN AND AMERICAN WRISTWATCHES

ローレックス在庫も豊富。
御要望は御手紙でどうぞ。

ROLEX SPECIALISTS

MELROSE AVE

LOS ANGELES

We invite you to visit our Melrose Avenue store and view our vast selection of beautifully restored wristwatches.

FAX OR WRITE FOR OUR VINTAGE ROLEX CATALOGUE

 WANNA BUY A WATCH?
(213) 653-0467
FAX: (213) 653-1405
7410 Melrose Avenue
Los Angeles, California 90046

BEAUTY...

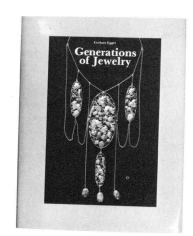

Generations of Jewelry
by Gerhart Egger

A beautiful history of the art form begins with a liberal discussion of fine jewelry's ancient history as exotic amulets and symbolic ornaments and proceeds to explain and profusely illustrate developing trends in jewelry. Beautifully illustrated in color and black and white with over 450 photographs.
Hardcover, 224 pages $50.00

and POWER

The Power of Jewelry
by Nancy Schiffer

Hundreds of stunning color photographs of magnificent jewelry and fascinating legends associated with all the different gem stones are combined to form a fresh, new approach to antique and modern jewelry. A visual feast with interesting and informative stories.
Hardcover, 256 pages $75.00

Schiffer Publishing Ltd.
1469 Morstein Road
West Chester, PA 19380

J&P TIMEPIECES

FOSSNER TIMEPIECES

*We Specialize in Buying and Selling
High Grade
Wrist Watches and Pocket Watches*

Call or Write: Ask for Jeff or Peter: Monday–Saturday 12–7 p.m.
(212) 980-1099 or (212) 249-2600. Other Times Jeff (212) 924-3933

J&P TIMEPIECES or FOSSNER TIMEPIECES
1059 Second Avenue, New York, New York 10022
FAX (212) 935-0339

THE WORLD OF TIME
from Schiffer

American Wristwatches: Five Decades of Style and Design

by Edward Faber and Stewart Unger with Ettagale Blauer

Over 600 color photographs illustrate the development of the American wristwatch from the early years of the Twentieth century until its decline. Full of information about the companies and people who provided the drive and innovation that signified the American contribution to wristwatch design and style.

Hardbound, 288 pages $79.95

European Pendulum Clocks

by Klaus Maurice and Peter Heuer

This volume studies, in chronological order, three types of large clocks: wall, cabinet and free-standing, with particular attention to clocks from France, England, Holland, Scandavia and the German-speaking countries. More than 400 beautiful B/W and color photographs, obtained from collectors and auction houses thorughout the world, accompany thorough explanations of each clock.

Hardcover, 248 pages $59.95

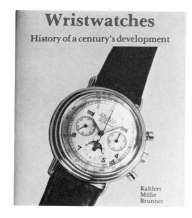

Wristwatches, History of a Century's Development

by Helmut Kahlert, Richard Muhe, Gisbert L. Brunner

This book displays, through 539 illustrated examples with identifying captions and supporting text, the fascinating world of extremely carefully calibrated moving parts, and changing technology and style. Wristwatches from around the world are considered in groups by their makers, technological changes, construction and automatic features.

Hardcover, 410 pages $59.95

Pocket Watches

by Reinhard Meis

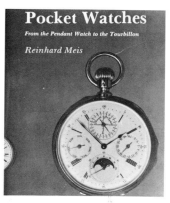

Historically and visually this is the most exciting volume on pocket watches that ever has been compiled. 915 photographs of old and newer styles show the development clearly, and the text explains the mechanisms and manufacturers.

Hardbound, 316 pages $50.00

SCHIFFER PUBLISHING LIMITED 1469 MORSTEIN ROAD WEST CHESTER, PA 19380

CHIUZAC GALLERY

This elegant gallery is world renowned for maintaining the highest quality inventory of vintage wrist watches, by makers such as Tiffany, Cartier, Patek Philippe, Rolex, Hamilton and Gruen. All are design oriented, chosen for their unique qualities and guaranteed.

The gallery's buyer is Jac Zagoory, author of A TIME TO WATCH, a book considered by many to have been a major factor in the rising popularity of collectible wrist watches.

510 MADISON AVENUE   NEW YORK, NEW YORK 10022
Tel: (212) 832-2233   Fax: (212) 319-7691

TIME WILL TELL
ANTIQUE WRISTWATCHES

962 Madison Avenue
(75th Street)
New York, New York 10021
(212) 861-2663

New York's finest collection. Rare and unusual vintage and classic wristwatches. Prized collectables, sound investments, fashionable and superb timepieces. Fully warranted. Expert repairs and restorations. Free catalog available.

**bought/sold
consignments accepted**